PRAISE FOR *SUPPLY CHAIN TRANSFORMATION*

Supply Chain Transformation is an engaging and essential read that I can confidently recommend to every supply chain leader. Michael Mowat combines academic rigor with practical insight, offering clear guidance on data-driven decision-making, automation and end-to-end visibility. With examples from leading companies including P&G, Unilever, Amazon, DHL and Tesco, he shows that even the best supply chains have room to evolve. It's a timely reminder that supply chain is a force for global stability and innovation. A must-read for those shaping the future of supply.
Graeme Carter, CSCO, Coty

In *Supply Chain Transformation*, Michael Mowat offers a rare blend of clear frameworks and hands-on playbooks. He shows how to move from visibility to action, redesign decision rights and align incentives across partners leveraging rapidly evolving digital technologies. This book serves as an important reminder that supply chain isn't a function; it's the engine of strategy and the key to executing it at pace.
Vineet Khanna, former Global Head of Supply Chain, Nestlé

This book is a much-needed resource for supply chain practitioners and consultants, as well as any academic planning to teach a course on digital supply chains for MBA, Executive MBA and other professional development courses. The chapters are well organized, covering drivers, benefits, opportunities and challenges, real-life examples and easy to understand tables and figures.
Atanu Chaudhuri, Professor, Technology and Operations Management, Durham University

A great read for the beginner and seasoned veteran alike. The history and real-world examples alone are worth a read. The real value is in Michael Mowat's ability to dive deep into what makes supply chains successful but also provide practical advice to take a supply chain to the next level.
Sandy Gosling, former Supply Chain Partner, McKinsey & Company

This is an excellent read, whether you are just starting your career or an experienced manager or leader. This book grapples with the supply chain challenges and how technologies can be adopted to drive real business change and competitive advantage. It is comprehensive, practical and covers everything from human technology interaction to sustainability and social responsibility.
Simon Finch, Supply Chain Director, Harrods

An essential read for professionals seeking a clear understanding of the digital revolution and of the opportunities it presents. Michael Mowat brilliantly builds on his extensive experience and knowledge to provide a rigorous and reliable guide, whether you're developing a digitization strategy or executing a digital transformation. A definitive reference work for supply chain executives.
Sami Naffakh, former CSCO, Reckitt and Arla Foods

Supply Chain Transformation offers a comprehensive review of the technical landscape in supply chain, current and future technologies and a bridge between implementation strategies and success KPIs. Critically, the book connects technology, process and people in defining future success. A good read for anyone in the industry stepping through our ever-changing world.
Gary Horsfield, CSCO, Bic Group

Michael Mowat's book *Supply Chain Transformation* is a welcome foundation to all those who are active in the supply chain, as well as the students of the supply chain. Mowat brings perspective and furnishes not only the concepts but the tools, processes and essential critical thinking. I certainly plan to leverage the learnings with my students and industry partners.
Joseph Agresta, Program Director, Master of Supply Chain Analytics (MSCA), and Assistant Professor, Rutgers Business School

Michael Mowat has written a very comprehensive book on supply chain transformation using technology. It is highly practical with expert insights and knowledge that can be immediately used, and includes various detailed overviews of all relevant supply chain technologies. I highly recommend reading it.
Ivo Böhm, CEO and Founder, B&O Supply Chain Ventures

Striking that rare and effective balance between theory and application, Michael Mowat's *Supply Chain Transformation* provides a critical evaluation of the role and effectiveness of technology in modern supply chain management. Examining supply chain technology in all its guises, as well as providing a framework for effective deployment and evaluation, *Supply Chain Transformation* is a must-read for every supply chain practitioner and leader.
Jonathon P Karelse, Co-Founder and CEO, NorthFind Management

In today's complex and uncertain world, there has never been a greater need for a supply chain leader to ensure their supply chain performs for today and transforms for tomorrow. Backed by useful real-world examples and references bridging theory to execution, *Supply Chain Transformation* provides a thought-provoking and comprehensive resource for the supply chain and enterprise leader as they look to navigate the challenges that the modern supply chain is presented with. Highly recommended reading.
Chris Yong, Global Senior Director and Head of Supply Chain, International Operated Markets Business Unit, McDonald's

This book examines a plethora of technologies as the key enablers for providing a competitive advantage and explores why they are a major catalyst for supply chain transformation. Other topics covered include the business case for investing in these technologies and the role of human factors. This book will be invaluable to supply chain professionals and university students aiming to enhance their knowledge regarding the transformative role of technologies in modern supply chains.
Michael Bourlakis, Director, Centre for Logistics, Procurement and Supply Chain Management, Cranfield School of Management

Supply Chain Transformation

How to leverage technology for competitive advantage

Michael Mowat

KoganPage

First published in Great Britain and the United States in 2026 by Kogan Page Limited.

Kogan Page
Kogan Page Ltd, 2nd Floor, 45 Gee Street, London EC1V 3RS, United Kingdom
Kogan Page Inc, 8 W 38th Street, Suite 902, New York, NY 10018, USA
www.koganpage.com

EU Representative (GPSR)
Authorised Rep Compliance Ltd, Ground Floor, 71 Baggot Street Lower, Dublin D02 P593, Ireland
www.arccompliance.com

Kogan Page books are printed on paper from sustainable forests.

ISBNs
Hardback 978 1 3986 2337 8
Paperback 978 1 3986 2335 4
Ebook 978 1 3986 2336 1

British Library Cataloguing-in-Publication Data
A CIP record for this book is available from the British Library.

Library of Congress Control Number
2025039344

Typeset by Integra Software Services, Pondicherry
Printed and bound by CPI Group (UK) Ltd, Croydon CR0 4YY

CONTENTS

PART TWO Digital technologies for supply chain advantage

LIST OF FIGURES AND TABLES

Figures

Tables

ABOUT THE AUTHOR

Michael Mowat is a distinguished leader in supply chain management, with more than 25 years of comprehensive experience spanning both industry and consultancy. Widely regarded as an authority in his field, he frequently shares his expertise through published works and speaking engagements.

He is a Fellow of the Chartered Institute of Logistics and Transport and has served on the Innovation and Sensing Committee for the Association for Supply Chain Management. His academic credentials include an MSc in Operations Management, an MBA and certified Supply Chain practitioner.

Throughout his career, Michael has operated on a global scale, supporting organizations in navigating complex supply chain challenges and implementing effective solutions. He remains actively engaged in advancing the discipline by mentoring future professionals, advocating innovative strategies and delivering guest lectures.

PREFACE

The motivation for writing this book came from what I was seeing and reading in regard to the role of technology in supply chain management. Many efforts from companies over the last ten years – grappling with legacy systems, trying their hardest to customize and manually intervening to circumvent technology to run the supply chain – were becoming very painful to watch. I felt the need to step back, consider how we should look at the role of technology and do things differently to achieve the desired outcomes we all want.

As I was finishing my undergraduate degree, where the decision of following a path to being an Accountant didn't resonate with me, I chose to do an elective in Operations Management which was an inflection point. It came naturally to me, and I have never looked back since. Whilst undertaking my first master's degree, I secured a job with a US Consumer Goods company who had operations at the time just outside Glasgow, Scotland (as well as many other sites globally). With them I was introduced to the AS400 ERP system; in their case it was coming to its end of life but quickly learning all the shortcuts about how to run it from a supply chain perspective became a joy for me. During those first few years, I worked with many talented people who taught me so much about the supply chain and technology which aided me greatly as I finished my initial studies and transitioned into working life.

Early on, I was exposed to materials management which provided me with so many fond memories, from having to search trailers for tomato paste, to running around production plants trying to understand why a batch was taking so long, to negotiating quantity sizes on labels due to confusion over a label SKU code, to having to quickly change into a suit because a customer was coming to do an audit. From there I was able to move through a variety of roles including inventory management, demand planning, manufacturing, logistics, procurement, joint venture set-ups, cost reduction and managing supply chain functions.

The next stage of my career began when I entered the world of consulting where I was exposed to a variety of sectors and many different company problems. I became interested in strategy, how the supply chain is stitched together beyond the physical, and the advancements of technology. Over the years I have had the opportunity to travel extensively with work across Europe, Africa, the Middle East, Asia and North America, making lifelong friends along the way.

As I write this, a few stats of my supply chain life to date include working with 10 of the top 25 supply chains in the world (if you believe a certain analyst's yearly view!), having been in over 300 warehouses and in 50 different forms of transport, seeing the rise and fall of a few companies, and being front row on two of the world's biggest supply chain challenges. (I should just say I was asked to be part of sorting them out rather than the cause!)

The evolution of supply chain technology over my working career to date has been quite remarkable and shines a bright light on the gaps systems in all their different shapes and sizes have left. One of my underlying gripes today is how much so many of these supply chain technologies promise to offer and fail to; you just have to read about them claiming to have AI capability when in fact if you look under the bonnet, there is just good old solid mathematics. My wish is that that changes over time – it is changing as this book goes to press in late 2025 but not at the rate that many claim.

This book takes a deep dive into supply chain technology, considering the different capabilities that are now available to organizations. It then pivots to how to go about building the case for change, considering the ramifications and impacts on people, before looking to the future and what further advancements are beginning to appear to aid companies in the drive for competitive advantage in their supply chains.

Today, alongside my day job, I speak at conferences and guest lecture. I think I get more out of the lectures than the students, enjoying their questions and energy, and hopefully I will have done a small part in getting a few of them excited about devoting some of their career to this great profession.

Writing this book has allowed me time to reflect and has provided perspective on what I should have maybe done differently in certain situations – but if there is one thing I hold onto, it is that I love understanding and helping to improve supply chains in any guise.

This book is based on my experiences and those generous to share with me theirs. Any opinions I provide are mine and have nothing to do with the views of companies I have worked for or do today, so please judge it on me alone.

PART ONE
Introduction to the supply chain dilemma

Supply chain management

01

Opportunities in the digital age

This chapter will provide an overview of the current supply chain landscape, the challenges of staying competitive and the opportunities presented by digital technologies.

CHAPTER AIMS

- Grasp the complexities of the modern supply chain and its impact on competitiveness.
- Pinpoint areas for improvement and opportunities for growth in the digital age.
- Appreciate the initial link between technology and driving competitive advantage.

'Supply chains compete, not companies.'

MARTIN CHRISTOPHER[1]

The term 'supply chain management' was first coined by Keith Oliver in a 1982 interview with the *Financial Times*.[2] Oliver, a pioneer in the field, described supply chain management as the process of:

- **Planning:** Coordinating and strategizing supply chain operations.
- **Implementing:** Executing supply chain plans and processes.
- **Controlling:** Monitoring and adjusting supply chain operations to achieve objectives.

The primary objective of supply chain management, as defined by Oliver, is to satisfy customer requirements as efficiently as possible. This involves the movement and storage of raw materials / packaging, work-in-progress inventory and finished goods from origin to consumption. The fact that the term has remained relevant for over 40 years is a testament to its simplicity, encompassing the broad range of activities involved and highlighting its importance as a concept of modern business.

While this was a watershed moment, looking back over history, there have been many moments to celebrate of supply chains in action. In the 19th century, Alexander Walker, the son of John 'Johnnie' Walker, found ways to put whisky in the hands of customers, from the US to Australia, via a complex network of distributors, focusing solely on product quality and receiving payment for his efforts. Fortunately for him and others during this era, if you were in the market with products that had long shelf lives (it could be argued that whisky never goes off!), a journey of months wouldn't matter.[3] Most supply chains pre-20th century were local, managing at best to reach regional levels, but the introduction of railroads then made it possible to provide consistent products on a regular basis.

Through the 20th century, supply chains continued to expand, moving from local to global markets with a focus on optimizing manual processes, helped by the introduction of mechanics in various parts of the supply chain such as in manufacturing and logistics. Analytics began to be used in the supply chain post Second World War by the military to aid with logistical elements. Unit loads (i.e. a group of items stored or transported) were developed to support this effort and their introduction became a significant breakthrough. Companies experimented with different forms of these to improve handling of products, leading to the introduction of pallets and containers. As Marc Levinson (the prominent financial journalist) suggests in his book *The Box*,[4] there was no 'aha' moment, but different approaches tested in North America and Europe were adopted as new ways of working. The real innovation came from deregulating transport and the integration of East Asia into the world economy, which opened the opportunity of where and how manufacturing would be done in the world moving forward.

Distribution started to take off with the emergence of logistics providers (e.g. DHL, FedEx) in the 1960s and 70s. They offered capabilities to receive, store and ship products across the globe via different types of freight transport that responded to the time dependent criticality. At this point, the term 'physical distribution' became part of the vernacular.

The supply chain was beginning to be exposed to innovative ways of working, in terms of physical distribution and the emergence of technology as well. In 1963, NASA was struggling to track parts in its space programme and urgently required a solution. IBM's Uri Berman with Rockwell's Peter

Nordyke went about assembling a team. It took lots of experimenting but in 1967, the solution was ready to ship: the first computerized inventory management and forecasting system called Information Management System.[5] The complexity of operations in companies was becoming too much for only manual ways of working and solutions involving technology were starting to be experimented with.

At a similar time, the concept of material requirements planning (MRP) was forming to help companies deal with the complexity of manufacturing, as there were not consistent approaches to managing production and inventory. In 1964, Joseph Orlicky, a manufacturing specialist at IBM implemented the first MRP system at Black & Decker with the aid of an IBM 1401 computer.[6] This was a revolutionary step in how to have materials available for production, while keeping inventory as low as possible at the product and finished goods level. Planning activities, deliveries and purchasing all were working somewhat in concert.

Then another breakthrough came with electronic data interchange (EDI), introduced in the supply chain by Ed Guilbert, an army master sergeant in the US Army, in 1968. He developed a way to communicate electronically to aid shipping for the US Army which came from his original idea in 1948 of a manual structured shipping manifest.[7] This became popular in the 1990s as a form to share information between parties engaging in supply chain movement of goods.[8]

In the late 1960s, the first automated storage and retrieval system (ASRS) was developed by Daifuku in Japan.[9] Although the concept of storage and retrieval was developed in the 1950s by Demag, it was manual in operation when implemented into the Bertelsmann book club warehouse in 1962.[10] ASRS solution, originally used for heavy pallet loads, has evolved over the years into various versions that can handle products of all different sizes, storing them in an automated system controlled by software. So, the 1960s should be marked as the decade when technology, still in its infancy on many fronts, was being used to aid companies to improve their operations.

The development of technological solutions for different elements of the supply chain continued with the introduction of the barcode in 1974, allowing products to be scanned.[11] This was very helpful for tracking inventory across the supply chain, and the fact it is still in use today shows what a marvellous invention it was. The development of the first warehouse management system (WMS) by now-defunct company Logisticon came in 1975, which enabled it to update inventory records in near real time and reduced time to understand and count how much inventory it had.[12] I should state here that the definition 'near real time' can be interpreted differently depending on the individual: it can mean immediately or it can mean within an agreed timeline. In the example above, near real time meant that as soon as an item was scanned, it would write back to the system within an hour; today it could be instantly.

As personal computers entered the arena in the early 1980s, technology started to become part of daily business life. It helped supply chain professionals improve how they completed their jobs, with spreadsheets becoming the go-to option for planning work. MRP became manufacturing resource planning (MRP II) during this decade, expanding the scope of the original MRP with functionality to address capacity planning, shop floor and production scheduling. The step change was to expand the scope to integrate manufacturing with financial and operational planning.[13] This shows the experimentation that was being done with supply chain technology to help solve this problem and look to where next we could use it.

Transportation modules were developed as a separate model, and this grew by adopting EDI within these systems. SAP in 1987 was first to develop a solution, quickly followed by companies such as Descartes and Global Logistics Technologies that provided users with a view of where shipments were without having to be linked to other solutions (i.e. standalone).[14]

Into the 1990s, technology was now becoming of interest to all companies regardless of size, as it could aid employees to drive efficiency and improve decision-making. All parts of the supply chain were ripe for technology with MRP II evolving into enterprise resource planning (ERP), which moved beyond manufacturing processes into encompassing finance, human resources and other elements of supply chain management. Manufacturing execution system (MES) emerged, defined as a 'dynamic information system that drives effective execution of manufacturing operations'.[15] There were two early versions: one that was standard and the other that was customizable.

Concurrently, advanced planning and scheduling (APS) systems were being developed, which focused on planning, programming and execution of supply chain operations. They were addressing the need to optimize production plans considering constraints such as raw material availability, capacity of production lines and service levels.[16]

In warehousing, there were other breakthroughs. In 1996, J Edward Colgate and Michael Pershkin, professors at Northwestern University, responded to an initiative by General Motors and invented the first collaborative robot (cobot) that could be used safely side by side with humans.[17] Also during this period, the emergence of the internet meant the way companies were servicing customers was rapidly changing. Amazon was at the forefront of this as the leading internet retailer, going public under two years from its creation in 1995.[18] The impact of Amazon on supply chain evolution should not be underestimated because here was a company that was able to ride the ups and downs of being in retail and survive even when all eyes were focused on profits. It doubled down on investments to improve the customer experience, which was predominantly based on how it could make the supply chain operate with flexibility and speed. We will revisit Amazon a few times during this book.

The next significant milestone for technology in supply chain was Industry 4.0 in the 2010s,[19] which started to offer the promise of artificial intelligence (AI), Internet of Things (IoT) and data. These technologies helped companies disrupt the way they were using existing technology in the supply chain and we will explore these more in later chapters.

In 2020, when the Covid-19 pandemic occurred, the importance of supply chains came to the forefront.[20] Initially, they came to a sudden stop and then had to reassess demand and supply minute by minute as markets opened, closed and then opened again. The event also challenged how supply chains are set up and led to many reviewing how to improve visibility and resilience, both themes that will be discussed in more detail in Chapter 4.

When we think about the building blocks of a supply chain from physical and technology perspectives, we can see how they must execute the vision of companies to serve customers in concert. How the supply chain is set up and operates will aid companies not just in satisfying their customers but can be the reason why they are chosen to provide goods in the first place and provide them with competitive advantage. Interestingly, it took time for us to innovate with technology but once we overcame that first hurdle of using it, it quickly became part of the solution for all elements that make up modern supply chains, typically led by those supply chains tied to large global companies.

Figure 1.1 is generic in nature because supply chains need to be considered through the lens of a specific company. The company may be a retailer who is interested in the inputs of movement to its part of the supply chain, which is mostly in distribution centres and stores. They provide signals to

Figure 1.1 Overview of physical and technology view of a supply chain

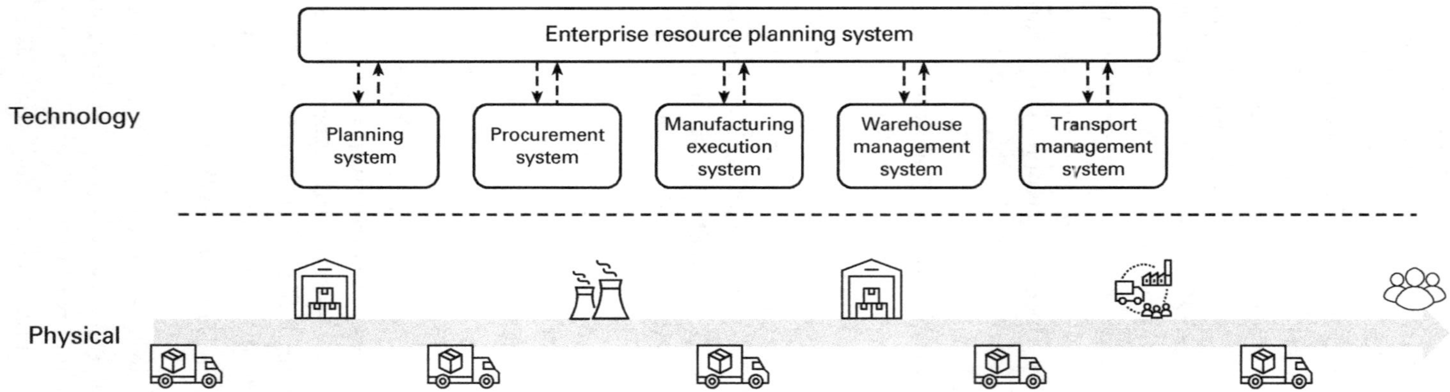

partners upstream on what their requirements are based on their understanding of what customers want and inventory levels across their estate. If looking at a consumer products company (CPG), they are focused on transforming raw materials and packaging inputs into a finished goods output, which they can then sell to, in most cases, the distributor / retailer (they are one or more steps removed from the customer typically). They will be very interested in the final customer and will incur expenditure understanding what the customer wants and how to make a product that appeals to them. Therefore, they will be driven to capture that demand or create new demand. This is similar for companies that deal in industrial products such as those aligned to providing an input to a CPG or others in a business-to-business set-up. At a high level, companies that provide goods will have a similar supply chain set-up, with subtle differences in the way they gain inputs, transform them and then output downstream.

Today, modern supply chains are global in nature. If the company's own operations are not global, then those upstream and downstream will most likely be. The supply chain is complex and is made up of interconnected networks of activities that succeed or fail based on the combination of people, process, technology and data to produce and deliver products or services. It is rarely a static operation, with constant changes of demand affecting it and how it has to respond. The barriers to service customers have reduced significantly, with customer expectations also challenging the supply chain, which was originally set up to be static in nature, but following a standardized operandum will result in losing competitive advantage. Producing at volume can be difficult to achieve now for companies that, in the past, looked to economies of scale to win. Also changes to regulatory requirements can mean changes are required at a country, region or global level for different products. The global markets that supply chains service are inherently exposed to factors that make predictability and stability a challenge, which supply chains must adapt to.

Introduction to innovation

Technology innovations proliferate over time, which is continuously changing supply chains. Companies must determine which innovation will help manage their cost base better, provide an advantage in the market and help their employees do their roles more efficiently. Many of these we will discuss in detail in this book. Currently, the role of the traditional systems is being challenged with new versions or different systems that encompass more elements of the supply chain or provide greater advancements.

These innovations are exciting but can also be daunting because they assume companies are, in many cases, at a mature operating level. The challenge for modern organizations is to implement and iterate the interdependence of people, process, data and technology to achieve success within the supply chain context, based on Leavitt's Diamond model from 1965,[21] which demonstrated the interdependences of people, process, data and structure.

Figure 1.2 below showcases the four elements that will continue to appear throughout this book. Focusing heavily on one comes with consequences on the others, which can be to the detriment of the company and its goals overall.

Sadly, many research providers' opinions in relation to these innovations can be over-inflated due to the unhealthy relationship that can exist between them, technology providers and their customers. When you hear of a leading CPG in the supply chain field, you are initially drawn to it, only to understand with research that it is using that capability in one market but can't scale it because the business case was either too narrow in focus to address that market's issues or did not provide the perceived value once implemented

Figure 1.2 People, process, data and technology

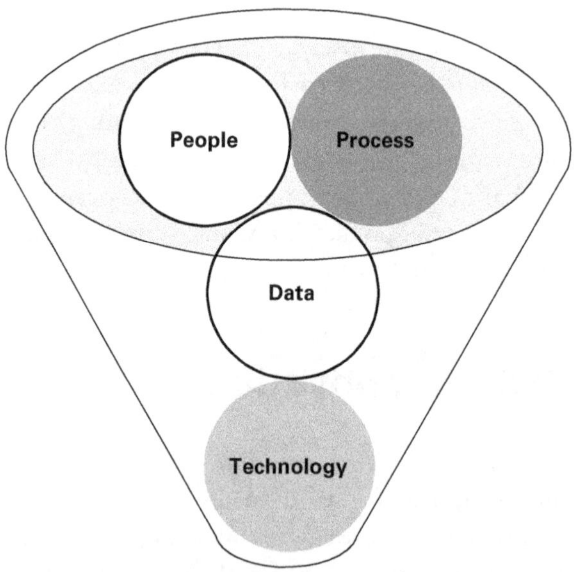

SOURCE Adapted from Leavitt's Diamond model (1965)

in one market, or that the pilot failed, or that forcing a global approach on a local market may be adding too much complexity. The reality is that not every technology implementation will be successful, but experimenting and trying to improve the supply chain is the right approach. Having a supply chain that has many workarounds, manual activities that are not focused on value adding, partial use of technology capabilities and heavy spreadsheet use is a 'plaster' mentality. It might be appropriate in the short term but ultimately points to the need for 'surgery' which has been an unpopular choice due to it requiring the investment of money, time and people. However, companies sticking to what they know best had led to significant underinvestment in supply chain capabilities pre-Covid, where many companies prioritized rewarding investors with dividends or buyback of shares.[22] Since Covid, there has been more appetite again to invest aligned with an appreciation of the supply chains' importance to how a company performs.[23]

Addressing sustainability

An important theme that modern supply chains must address is sustainability, considering both environmental and social responsibility. With the global nature of supply chains, there is an impact on all companies that form it. Being able to understand environmental impact has been front of mind of companies for over 20 years. In the late 1990s, the World Resources Institute (WRI) and The World Business Council for Sustainable Development (WBCSD) came together to find a standard way to account for greenhouse gases. This led to the creation of the Greenhouse Gas (GHG) Protocol, which is an organization that issues standards on climate. Since its first corporate standard published in 2021,[24] it has developed frameworks and standards to assist companies on reporting, which have fallen under the main headers of scope 1, scope 2 and scope 3. By 2023, 97 per cent of S&P 500 companies reported on their emissions using GHG Protocol.[25]

- **Scope 1** is direct GHG emissions that occur from the work the company does within its own business transforming raw materials / packaging into finished goods.
- **Scope 2** is indirect GHG emissions that occur through the purchase of energy to run the business.
- **Scope 3** is GHG emissions that do not occur in the company's business; therefore these are not incurred directly but are within the up or down value chain associated with the company's activities.[26]

Figure 1.3 Understanding scope 1, 2 and 3

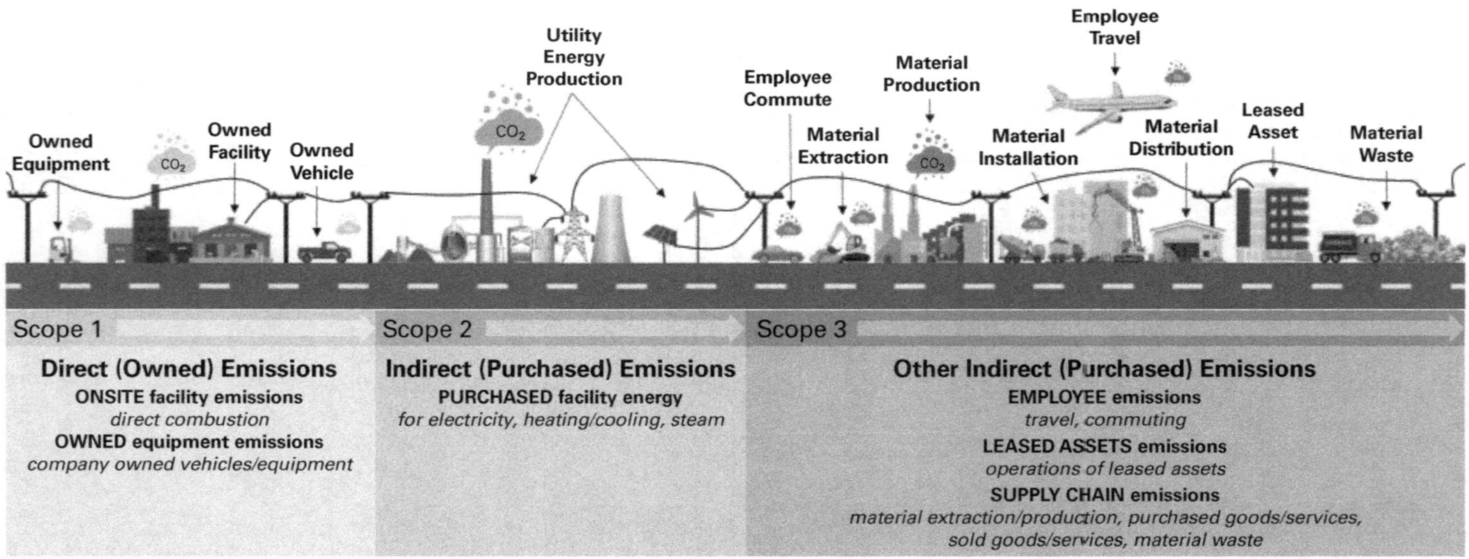

SOURCE Reproduced with kind permission of Stacy Smedley

The purpose of each scope through a company lens is to provide guidance on reporting emissions, help to assess the impact of a company's value chain, and identify reduction initiatives and product standards. This should be coupled with International Organization for Standardization (ISO) standards. ISO14064 is the main one that guides companies in reporting, setting and achieving reductions of unsustainable practices and is applicable to all companies. As supply chain leaders evaluate how they are performing, decisions in relation to how they run the supply chain must be mindful of the impact this will have and encourage modernization, since technology is an enabler of efficiency in its truest form (or should be). Also, the ability to control different elements and/or have visibility over the supply base within global supply chains is key for monitoring and addressing sustainability. There are sadly cases where there has been a breakdown in responsibility to society. We normally hear of these cases when a company has been exposed for unsustainable and unethical activities. Forced labour and child labour are significant issues within the supply chain. Although this has improved greatly over the years, likely owing to the improved regulation and the corporate social responsibility (CSR) programmes that are well supported and encouraged within companies, and company self-interest to maintain brand reputation, it is still a pertinent issue to be aware of. We will pick this up again in Chapter 8 because technology can significantly aid companies in addressing it.

Minimizing risk and uncertainty

Risk and uncertainty have always been a concern for supply chain leaders; being able to mitigate risk with the additional costs of protecting the supply chain can be difficult. There are many local and global risks that can cause disruption. Since 2020, supply chain leaders have been dealing with Covid, conflicts in Europe and the Middle East, tariff announcements, countries struggling to cope with inflation, high interest rates and natural disasters. From a supply chain perspective, the need to have control over the movement of goods, where to source from, where to have manufacturing facilities and the overall strategy to deliver against this tumultuous backdrop requires decisions that might require years to implement. Technology can help companies make informed decisions, but uncertainty requires trade-offs. For example, in 2025, the US government stated its desire for onshore manufacturing, which could take many years for companies to adapt to. In some instances, onshore manufacturing could just require moving production lines to a current facility, but in many instances significant investments in new plant and production lines may be required. Then, workers must be trained. However, once this has been completed, the labour cost of US employees versus offshore countries, such as those in East Asia, may not be comparable (i.e. a US employee cost will most likely be higher), which will lead to higher prices for consumers. Also, as stated by Apple, the advancements in skill set could be generationally different as well, resulting in the need to move production and a subsequent loss of productivity. Therefore, the potential issues could go beyond just labour costs in the future.[27]

Staying competitive requires companies to face these challenges, while adhering to standards. Managing complexity requires supply chain leaders to find a way to be efficient, agile and resilient simultaneously. Visibility, the holy grail of the supply chain, is difficult to achieve in just your part of the supply chain, let alone achieving this upstream and downstream. What level of visibility is needed and appropriate must be agreed because there are only so many insights that can be gained with visibility. It is fair to say that having near real time insights in your end-to-end supply chain helps you make better-informed decisions. However, if you're not careful, you could end up reacting to every bit of information or be paralyzed with analysis and will not be able to make decisions. Establishing and enhancing collaboration in your supply chain is very advantageous because parties will be able to share information and get into problem-solving much quicker when issues arise. The difficulty comes from how to foster this at speed and how companies engage with each other and establish trust, so they work aligned to a common goal, can address risk and in a sense 'see around corners' to keep supply chains working effectively.

Technology provides opportunities to assist supply chain leaders as they tackle unpredictable times and support companies with gaining competitive advantage and growth. There are four themes that are crucial to consider and will be explored in more detail in this book:

- **Visibility:** Technology that assists companies with being able to see what is happening in their supply chain and the wider ecosystem they operate, as well as establish trust with relevant stakeholders from source to the customers.
- **Simulation:** Being able to look at ways to optimize and simulate scenarios to support decision-making.
- **Analytics:** Leveraging data, not to just see what is going on in real time, but also to anticipate and respond to changes in demand, supply and market conditions.
- **Automation and robotics:** Increasing efficiency and productivity within the supply chain by deploying capabilities that can assist employees with their activities.

While technology has been seen to aid companies in how they operate the supply chain, it would be naïve to think it has all been great news. For example, significant promise has not been delivered in consistent productivity gains. Robert Solow, the renowned US economist, famously articulated this problem in 1987: 'you can see the computer age everywhere but in the productivity statistics.'[28] A productivity surge was seen seven years after this statement, with George Graetz and Guy Michaels finding that technology specific to deploying robotics within operations were helping drive improvements in productivity between 1993–2007 based on their research.[29] However, in 2023, Gartner presented a graph that showed in the eight years prior to 2023, productivity in relation to supply chain was not budging,[30] which suggested companies were struggling to squeeze out gains from technology in the supply chain. Furthermore Dave Clark (ex-Chief Supply Chain Officer at Amazon) summed up his view of supply chains as 'inefficient, chaotic, costly, suboptimal and completely unfit for the modern world'.[31]

This book will explore this, reset expectations and challenge the reader to think about what they can do differently to gain the perceived benefits of technology within the supply chain.

The chief supply chain officer

I wanted to end Chapter 1 mentioning about the emergence of the title of chief supply chain officer (CSCO), which has only been in existence since the turn of the century. To put it in context, by 2012, only 75 companies in the Standard & Poor 500 index had one titled as such.[32] Prior to then, there were people in senior roles within supply chain, but they may have fallen under the remit of chief operating officer or chief financial officer, for example.

Defining the role in its simplest terms, a CSCO is responsible for 'developing and executing the supply chain's long-term strategic plan – ensuring it is appropriately future-focused'.[33] I would add that the role involves being the steward of the supply chain to meet customers' expectations at a cost that is sustainable for the company to be successful.

The role requires a broad skillset that can jump between strategic and day-to-day operations depending on various situations they can find themselves in. The best CSCOs that I have worked with demonstrate excellent communication skills to engage and influence different stakeholders' groups, ability to motivate and energize colleagues and deliver supply chain performance that consistently helps the company succeed. This is by no means an exhaustive list, but relevant people they would engage with include supply chain personnel across sub-functions, commercial, finance, IT, HR, board of directors, investors, customers, suppliers, trade unions and associations. When considering technology, the CSCO must be the requester of a new implementation, not the receiver of the request, which is still a contentious issue in many organizations since many decisions may be led by important system legacy upgrades. Research conducted examined the role and concluded it is here to stay and brings value when operating in a volatile business environment.[34] We will look at this further in Chapter 3 and in Chapter 12.

Conclusion

In this first chapter, we have gone back in time to understand some of the major milestones in the evolution of supply chain and technology that underpins it. It is incredible to see how much technology has changed over the course of history, how it has enabled supply chains to become global and connected and how it continues to promise more.

It is important to understand the challenges CSCOs and supply chain leaders face today so we can be better prepared for the future, know what the priorities should be and how technology supports this agenda. Supply chains do face more uncertainty than in the past but also have more advanced capabilities to assist them in being able to bend and bounce to respond.

We have briefly framed the opportunity technology can bring and how that supports companies in gaining and maintaining competitive advantage. The linkage between technology and this advantage falls under various themes, including visibility, agility, efficiency and being a driver of innovation, which will all be addressed in detail throughout the course of this book.

Endnotes

1. Martin, C (1992) *Logistics and Supply Chain Management: Strategies for Reducing Costs and Improving Services*, Pitman Publishing, London
2. Kransdorff, A (1982) High stock levels – not the answer to volatile demand, *Financial Times*, 4 June
3. Morgan, N (2020) *A Long Stride: The Story of the World's No.1 Scotch Whisky*, Canongate Books

4 Levinson, M (2016) *The Box: How the Shipping Container Made the World Smaller and the World Economy Bigger*, Princeton University Press

5 IBM (2019) Information management systems, IBM, www.ibm.com/history/information-management-system (archived at https://perma.cc/G27S-5XY8)

6 Orlicky, J (1969) *The Successful Computer System – Its Planning, Development and Management in a Business Enterprise*, McGraw-Hill

7 Skinner, N (2018) EDI – 50 years old and getting stronger, Opentext Blogs, https://blogs.opentext.com/edi-50-years-old-getting-stronger (archived at https://perma.cc/QXH8-WYAM)

8 EDI Academy (2015) EDI development and establishment, EDI Academy, https://ediacademy.com/blog/edi-developement-stages-of-edi (archived at https://perma.cc/AG5N-JRCL)

9 Daifuku (2020) History of Automated Warehouses at Daifuku, Daifuku, www.daifuku.com/solution/technology/automatedwarehouse (archived at https://perma.cc/F3CF-VBUW)

10 Site Editor (2020) When was the ASRS system invented? Nova Racking, www.novaracking.com/When-was-the-ASRS-system-invented-id3513603.html (archived at https://perma.cc/GJ65-HMFS)

11 Weightman, G (2015) The history of the bar code, *Smithsonian Magazine*, www.smithsonianmag.com/innovation/history-bar-code-180956704 (archived at https://perma.cc/CEJ7-HUU4)

12 Armenta, A (2022) History of warehouse management systems (WMS), Control.com, https://control.com/technical-articles/history-of-warehouse-management-systems-wms (archived at https://perma.cc/5QPT-YXXJ)

13 Meersseman, T (2024) The evolution of material requirements planning: Tracing the history of MRP, The Grain, https://blog.thegrain.pro/en/evolution_of_mrp (archived at https://perma.cc/8QPH-7HG6)

14 Leichter, R (2024) History of transport management systems by decade: 1970s to 2020s, Cargoson, www.cargoson.com/en/blog/transport-management-system-tms-history (archived at https://perma.cc/4LXC-BPUS)

15 Applied Control Engineering (2020) The evolution of MES: From the monolithic systems of the past to the agile cloud-based systems of tomorrow, Applied Control Engineering, https://f.hubspotusercontent00.net/hubfs/5836655/White%20Papers/Evolution%20of%20MES/05_20_ACE_MES_white_paper_final.pdf (archived at https://perma.cc/Y6RG-NA7E)

16 Chern, C C and Yang, C C (2011) A heuristic master planning algorithm for supply chains that consider substitutions and commonalities, *Expert Systems with Applications*, 38

17 Northwestern University Press Release (1996) Cobot: A kinder, gentler robot, Northwestern University, https://peshkin.mech.northwestern.edu/cobot/pressrels96/pr1196.html (archived at https://perma.cc/SH4V-4G3V)

18 WSJ Writer (2024) On this day in 1997: Amazon goes public on the Nasdaq, *The Wall Street Journal*, www.wsj.com/livecoverage/cpi-report-today-inflation-stock-market-05-15-2024/card/on-this-day-in-1997-amazon-goes-public-on-the-nasdaq-rHJiDBcBc9Mnq5vg2JKZ (archived at https://perma.cc/42SW-KJXH)

19 Vaidya, S, Ambad, P and Bhosle, S (2018) Industry 4.0 – a glimpse, *Procedia Manufacturing*, 20, www.sciencedirect.com/science/article/pii/S2351978918300672 (archived at https://perma.cc/BQA6-AZ4N)

20 Ownes, N (2025) How COVID-19 changed US manufacturing: 5 years later, Supply Chain Dive, www.supplychaindive.com/news/covid-19-pandemic-impact-labor-supply-chain-five-year-anniversary/744684 (archived at https://perma.cc/2XYL-8Z48)

21 March, J (1965) *Handbook of Organizations, Chapter titled – Applied Organizational Change in Industry: Structural, Technological and Humanistic Approaches*, Routledge

22 Letzing, J (2021) Why a focus on shareholders may be causing supply chain backups, World Economic Forum, www.weforum.org/stories/2021/10/is-an-eagerness-to-please-shareholders-contributing-to-supply-chain-misery (archived at https://perma.cc/P8N8-WKAJ)

23 Requis (2020) Why investors are looking at supply chain technologies in 2020, Requis, https://requis.com/blog/why-investors-looking-at-supply-chain-technologies-2020 (archived at https://perma.cc/SYK8-A79Q)

24 Greenhouse Gas Protocol (2021) Corporate Standard, Greenhouse Gas Protocol, https://ghgprotocol.org/corporate-standard (archived at https://perma.cc/3G39-ARV9)

25 Greenhouse Gas Protocol (2023) In 2023, 97% of disclosing S&P 500 companies reported to CDP using GHG Protocol, Greenhouse Gas Protocol, https://ghgprotocol.org/#3 (archived at https://perma.cc/5A47-2BV2)

26 CarbonNeutral (2024) Scope 1, 2 and 3 emissions explained, Carbon Neutral, www.carbonneutral.com/news/scope-1-2-3-emissions-explained (archived at https://perma.cc/52GN-FLW5)

27 FT Reporters (2025) Why Trump can't build iPhones in the US, *Financial Times*, https://ig.ft.com/us-iphone (archived at https://perma.cc/7EHM-NP47)

28 Solow, R M (1987) We'd better watch out, *The New York Times*, 12 July, www.standupeconomist.com/pdf/misc/solow-computer-productivity.pdf (archived at https://perma.cc/DD95-B5VY)

29 Graetz, G and Michaels, G (2015) Robots at work, IZA Institute of Labor Economics, Discussion Paper No 8938, https://papers.ssrn.com/sol3/papers.cfm?abstract_id=2589780 (archived at https://perma.cc/5ZYR-L93J)

30 Lavelle, J (2023) Gartner says supply chain productivity declines cannot be solved by technology alone, Gartner, www.gartner.com/en/newsroom/press-releases/2023-06-05-gartner-says-supply-chain-productivity-declines-cannot-be-solved-by-technology-alone (archived at https://perma.cc/MR9X-2VWC)

31 Stat Times (2024) Dave Clark announces founding team of Auger, *Stat Times*, www.stattimes.com/logistics/dave-clark-announces-founding-team-of-auger-1353901 (archived at https://perma.cc/AP99-MZDA)

32 Roh, J, Krause, R and Swink, M (2016) The appointment of chief supply chain officers to top management teams: A contingency model of firm-level antecedents and consequences, *Journal of Operations Management*, Wiley Online Library, https://onlinelibrary.wiley.com/doi/abs/10.1016/j.jom.2016.05.001 (archived at https://perma.cc/Q899-LB86)

33 Gartner (2025) Chief supply chain officer leadership vision 2025, Gartner, www.gartner.com/en/supply-chain/trends/leadership-vision-chief-supply-chain-officer#:~:text=The%20CSCO%20is%20primarily%20responsible,the%20function%20are%20well%2Daligned (archived at https://perma.cc/7H6S-FXR5)

34 Sihvonen, J and Kauppi, K (2025) A chief supply chain officer matters – but only when you are in trouble! *International Journal of Operations and Production Management*, ISSN 0144-3577, www.emerald.com/insight/content/doi/10.1108/ijopm-11-2024-0985/full/html (archived at https://perma.cc/8LN9-38ZU)

The role of technology in supply chain transformation

02

This chapter will cover the role of technology in enabling supply chain transformation, including the importance of data-driven decision-making and the impact of emerging technologies.

CHAPTER AIMS

- Discover the transformative power of technology in supply chain management.
- Learn how data-driven decision-making drives business outcomes.
- Explore emerging technologies with potential for supply chain impact and innovation.

'Many supply chains are perfectly suited to the needs that the business had 20 years ago.'

JONATHAN BYRNES[1]

In Chapter 1, we read how the role of technology has evolved from the 1960s and how its purpose as an enabler in the supply chain, working in collaboration with people, process and data, can unlock benefits for companies. The key point is that it is an *enabler;* yes, it can automate tasks, it can do calculations quickly and minimize human errors, but we should be very cautious in thinking it is a 'silver bullet' that once implemented will lead to autonomous supply chains and we can enjoy long retirements. The adage of 'garbage in, garbage out' is still relevant, whatever the technology, hence the reason that it must work in tandem with people, process and data and that it is not more dominant than any others. Why it gets so much airtime is because we are driven by the excitement about the new solution: technology provides a pathway, which by the time we get on the path, our attention moves onto the next technology.[2]

What cannot be disputed is that companies need to evolve and those that don't will not prosper for long. There are enough cases of business demise seen around the world, blamed on unappealing products, slow routes to market, pricey innovation, failures of service and even illegal activities. However, the theme of changing market conditions is nothing new. If anything is new, it is the speed of change, being accelerated by the technological progress occurring in the world we live in today. Influences on the market are typically listed under the following categories, including:

- **Economic factors** – interest rates, inflation, GDP growth, currency fluctuations.
- **Geopolitical factors** – trade policies, global conflicts, elections and politics.
- **Social factors** – demographics, consumer behaviour, social media.
- **Environmental factors** – climate change, natural disasters, sustainability.
- **Technological factors** – digitalization, AI.

With all markets being affected by these factors, they now need to be served by a different supply chain: one that can live, breathe and prosper when it is challenged, as too often they are reactive to these factors with limited space to manoeuvre.

Economic factors

Economic factors have always affected supply chains and the markets they operate in and there is no hiding from them. Supply chains, especially those of a global nature, face difficulties day in and day out, which in many cases, they cannot control. This is where companies must make conscious decisions about markets they enter, the risks they take servicing customers in those markets and what this can mean for brand value and their growth prospects.

If we look at currency fluctuations, procurement engages the supply base to determine who will provide different raw materials / packaging and will, in most cases, secure volumes and pricing to negate risk of changes in currency. In some cases, if they operate a more mature procurement function, they may have multiple suppliers and based on factors like currency, can adjust preferences and orders more easily. Where challenges occur, is if the procurement team do not have long-term pricing in place because the price is fed through to cost the bill of materials (BoM) and the finished product. Therefore, any significant changes will have a material effect on the margin and the potential cost to the customer.

Fuel prices are another major cost driver that can be affected by economic factors (as well as others, such as geopolitical factors). If prices change adversely for the company, it can result in significant oncosts, which can again hit margins. For example, it could lead to the company looking at different methods of transport to keep the cost impact low, which could impact the environment factor by it shifting its mix of fleet to electric vehicles. However, that could be affected by interest rates since they may need to borrow monies to action that. These examples show the need for technology to support the supply chain, from having readily available information and the ability to run scenarios quickly to make informed decisions.

Geopolitical factors

From a geopolitical viewpoint, the risk of supply chain disruption is significant and therefore the need to be able to confidently understand what is going on and respond to it is critical. For example, in 2022, the UK and other countries imposed restrictions on Russian goods following its invasion of Ukraine.[3] In November 2023, we saw risk in servicing the Suez Canal region with a rebel group disrupting global trade;[4] and in 2025, tariff wars led by the US government.[5] Again, technology can aid companies in dealing with these challenges. Immediately, a solution that provides visibility comes to mind when facing these situations: there are many potential unknowns and implications that need information to be shared among parties to make decisions from. With technology, companies will be able to first understand what risks they have with shipments in transit and then be able to understand whether they need to do anything different such as re-route products, order more from a supplier or ship using alternative forms of transport. They will also be able to determine implications of tariffs on their pricing and costs and whether servicing the market is worth it. The complexity of operating a global supply chain comes into focus here because there are always going to be geopolitical risks at play, but the use of technology and data will help inform them of implications and recommend strategies.

Social factors

It is important to consider social factors, especially in relation to consumer behaviour as we have seen significant shifts in consumer expectations. Where consumers traditionally did much of the hard work by setting up high streets and shopping centres with stores that sell goods and expecting them to spend time browsing and purchasing, we are now living in a hybrid world where this traditional set-up is available (but declining),[6] as the use of technology and our comfort in using it has led to the emergence and growth of online shopping. When comparing offline and online shopping, we can see quickly that offline is constrained by physical availability and space. This might seem obvious, however, when you think of a supply chain and how it might be set up, the ability to service the consumer changes dramatically with what capabilities you need (e.g. physical assets, systems, inventory) and what level of service will capture the demand without incurring significant cost. Typical ways to service customers include:

- purchase in store, pick up in store
- purchase in store, home delivery
- purchase online, pick up in store
- purchase online, home delivery
- purchase online, pick up from warehouse
- purchase online, pick up from third-party asset (e.g. locker, another store).

Looking at these six examples of ways to service a customer, quickly you can see that there is complexity from a supply chain perspective: firstly in how to receive orders, then how to allocate inventory, then to execute delivery, which all have their own different costs. If we look solely at inventory, how does a planner know at any one time where their inventory is, what has been allocated, what is available and whether they need to order more from the manufacturer or whether to move the product to different locations? Orchestrating this would be very challenging without the use of technology providing information in near real time.

Also, consumers have been challenging retailers through order patterns because in this online world, where there are countless items to pick from, deciding can be difficult. For example, when choosing a dress, consumers must pick the design, size, colour and so on. This can lead to them ordering many more versions of dresses than they need, causing a high rate of returns on the shipped goods, which has both a financial and sustainability impact for the retailer. A recent report from the National Retail Federation (NRF) stated this issue of returns was now 16.9 per cent[7] however, others have argued it is 30 per cent,[8] but whatever the number, it is material. Companies have become wise to excessive returns and have put restrictions in place, de-incentivizing consumers by charging return costs. Also, consumers have the expectation of orders arriving within specific time frames, whether same-day, next-day, next-day timed or within 3–5 days. This has had a significant effect on how supply chains need to operate to satisfy consumers, which technology can enable.

Environmental factors

Environmental factors are another critical consideration for companies and their supply chains in the modern world. Consumers and governments have been prioritizing sustainability and practices for many years now, which global supply chains need to be part of the solution for. The impact of climate and weather alone is significant and we will focus here on these, while acknowledging there are many other parts to this topic that we will explore in more detail in later chapters.

Changes in the climate, such as rising temperatures[9] and changes in rainfall patterns[10] are already showing detrimental effects on agriculture, whereby securing a supply of commodities has changed dramatically. Natural disasters are occurring more regularly, leading to hurricanes, floods and wildfires that result in significant damage to supply chain assets such as factories, warehouses and ports. For example, Hurricane Florence[11] resulted in port closures for 11 days and some interstate roads closed for several days and disrupted how products move around to service customers. Here, technology enables companies to understand the impact their global supply chain is having on climate change: tracking this over time is a first step and is key for visibility. From there, companies can investigate different activities to lessen their environmental impact, like reducing the need for energy and water.

Technological factors

If we zoom in on technological factors, it is obvious to see how we have embraced technology as citizens of the world. Globally, 68 per cent of people can access the internet[12] and 78 per cent use a mobile phone daily.[13]

As technology continues to shape our lives and evolve, we are already seeing the importance of implementing it into our supply chains so that we can keep up, shifting from manual, siloed processes. Technology has fundamentally changed how supply chains need to be managed. Three elements that will be referred to again and again in this book when we consider next-generation supply chains are the increased use of automation[14] (e.g. AI and robotics), enhanced visibility[15] (e.g. near real time tracking of assets and goods) and improved decision-making[16] (e.g. using data and analytics).

Automation is being embraced to aid supply chain professionals across the supply chain. Warehouses, for example, have seen the introduction of both robotics and automated systems that have led to improved efficiency and changed the role of staff within the operation. There are many innovations in this space, including:

- how products are received inbound
- using scanners and systems to keep data live as soon as they come in from suppliers
- put away in storage via conveyor belts and automated storage systems
- warehouse management systems batching orders to create picking lists
- cobots aiding staff in picking and preparing orders to go out.

Enhanced visibility builds on this with the use of cloud systems that can be connected to share information in near real time to aid users in their roles. Technology, like IoT devices, helps continuously monitor and track, collecting and sharing data to systems to keep them updated. Near real time tracking is fast becoming a reality across different parts of the supply chain. For example, take a shipment from Asia of frozen goods: an IoT device can be attached to the goods to share relevant information along its journey, from being able to always report the temperature to make sure it adheres from a quality perspective, to always providing users with its location. This visibility helps reduce waste, stockouts and manage the customer's expectations.[17]

Improved decision-making will come from using technology to leverage data to drive insights. We will touch on analytics in Chapter 5, however, having insights from data also helps companies establish and monitor key performance indicators (KPIs) across the supply chain. Even today, there are many companies who have very basic ways of collating information regularly to report on these, but technology can help pull this information together quickly and visualize it to aid decision-making. By companies embracing these elements and keeping them front of mind when deciding where to invest in their supply chain, the opportunity to transform and innovate to be competitive will differentiate how successful they will be.

Digitalization and AI

Moving from traditional supply chain set-ups to transforming into digitalization supply chains has been bumpy, promising more than what it delivers in many cases,[18] e.g. the significant change in skills, technology and business strategy. The traditional approach has not changed, despite technological advancements: materials still flow downstream from suppliers through several steps to the customer, money flows upstream and information moves in both directions between parties along the supply chain. Technologies added as 'plasters' have helped to cover over the cracks of poor rigid technology solutions and shown promise but as pieces, they are only able to provide *some* value. It is only when you step back and look at the overarching view of the supply chain end-to-end, does it help you determine what you want to achieve with your supply chain and how technology can aid you in that journey.

The impact that technology has had already on upstream and downstream flows is worth considering but we need to look at what next can be done to make them even better. If we look at material flow, digitalization can help to move products (whether raw materials, packaging, work in progress, finished goods) quicker within a company and between companies in the supply chain from using automation equipment and making conscious decisions on forms of transport based on requirements of other players in the end-to-end supply chain. When looking at money flow, digitalization can help facilitate transfers and securing transactions through using cloud and blockchain, for example. However, the information flow is the most critical because based on this, the supply chain gains or loses competitive advantage.

'Knowledge is power' is a phrase that gets bandied around. Digitalization allows knowledge to be demarcated between parties that can help each of them succeed when they all know how to interpret and to act on it. Knowledge sharing without digitalization (e.g. opinions, gut instincts, what someone remembered last time they faced a similar situation) is inconsistent, lost in translation, subjective and not treated equally by all parties; therefore, the ability to have this in digital format, which is available at speed for all to use, is helpful for the ecosystem to succeed. Having this information codified using different digital capabilities to understand, assess and decide on outcomes is powerful. The traditional approach of working in silos and decision-making that require weeks to do, can now be done in no time. Workflows made up of many processes can be simplified and the ability to see where bottlenecks are occurring are other benefits that can be derived from digitalization.

Applying AI

AI is a significant topic and has come to the forefront since 2018. However, the term 'AI' was initially coined in a proposal for a meeting of some of the brightest scientists at Dartmouth College in 1956.[19] There are countless definitions for it, such as 'a field of science concerned with building computers and machines that can reason, learn, and act in such a way that would normally require human intelligence or that involves data whose scale exceeds what humans can analyse'[20] and 'technology that enables computers and machines to simulate human learning, comprehension, problem solving, decision making, creativity and autonomy'.[21] We have many ways we can apply and use AI-enabled technology in daily life now with digital assistants, navigation systems and wearables, for example.

Specific types of AI that are becoming popular in supply chain include the following:

- **Machine learning** – training algorithms on data to make predictions and optimize processes.
- **Generative AI** – taking data and creating something new from it.
- **Deep learning** – using neural networks (interconnected nodes / neurons in a layered structure) to analyse data and make predictions.
- **Natural language processing** – analysing and processing human language to improve communication and decision-making.
- **Computer vision** – analysing and processing visual data to improve inspection and detection.

All parts of the supply chain can prosper from the use of AI, and it revels in being fed data and asked to assist the user to make informed decisions at speed. Examples of this start with forecasting, a difficult mix of art and science. Scientifically, AI can assist by digesting and assessing multiple data sets, blending historical data, demand data (e.g. promos, revised formulations, positioning instore / online, manufacturing capacity) and seasonal data (e.g. macro factors, weather, local events) to help provide a more informed baseline forecast.

When completed, it can then hand over to the human, allowing them to focus on the 'art', which can include them assessing the forecast with their knowledge and engaging relevant stakeholders (e.g. customer, commercial, marketing, finance) to decide on what to propose for products. As they do this activity, they will also be considering inventory and finding ways to optimize this, which is also where AI can assist. For example, IoT sensors provide inputs of data and computer vision so the user knows inventory positions in near real time and can understand capacity constraints to optimize where the product needs to be. Furthermore, for balancing inventory to meet customer requirements, AI can be used to optimize routing and scheduling, for example. The role of AI with predictive maintenance, where machine learning and sensor data can aid decision makers on when to schedule maintenance work, reduces risk of machine and production line failures. AI also improves quality control, as computer vision and machine learning can inspect products and detect issues in near real time.

This all sounds great for supply chain professionals; however, with the benefits of AI come some challenges. Although we will discuss these later in the book, for now, it is worth mentioning a few of them briefly. AI needs data and with that comes the challenge of quality: making predictions and optimizing operations requires high-quality data, which many companies struggle to provide.

The ability to integrate systems and processes with AI is a further challenge because it needs to operate within the ecosystem and be plumbed into it (i.e. not be standalone). AI needs users with specialized skills and expertise, which are not skills that people operating in supply chain were traditionally required to have. However, this is changing and with the development of 'friendly' user interfaces, this, in time, should not be such a challenge.

Conclusion

We have covered significant ground here, skimming heavy topics that we will return to later, but the purpose is to frame the need for technology to improve how supply chains operate. What is becoming apparent is that supply chains need to evolve and leaders of them must embrace change and be open to finding ways to operate, serving a variety of customers through different channels. Economic, geopolitical, social, environmental and technological factors at play may change in importance; however, there will always be a time when something is affecting the global supply chain, so we must find ways to accommodate these as they appear.

The impact technology has played when reflecting on these factors is material in the context of supply chain. Technology's role in supply chain transformation is multifaceted: from being the reason for the change, to solving specific problems, to being the place to bring data together in a way that is easy to digest. The list goes on and on but when you bring it back to basics, it is there to help companies make decisions that drive outcomes quickly to help them be competitive. Digitalization of the supply chain is enabling processes that, when combined, create value. This has resulted in significant changes to skill sets and the roles required to operate within the supply chain world. Technology touches all elements of a supply chain and although its implementation can take time, as it advances, it will require companies to justify why they are not investing in it.

In this book, we will go into detail about how technology can help companies leverage analytics capabilities so that they can predict and respond to changes in demand, supply and markets, streamline workflows and the underlying processes, reduce costs, improve efficiency and provide flexibility to offer specific products and services to meet customer needs as they change over time. Robotics and automation can help with efficiency, cost and integrity of product as it moves through complex journeys from source to consumer. Sprinkle emerging technologies on that and you begin to imagine how feasible it is to enhance the supply chain further with AI. AI has become a significant topic in supply chain, with various applications emerging to improve efficiency, accuracy and speed. While AI offers these benefits, it also presents challenges, such as data quality, integration with existing systems and the need for specialized skills and expertise. Despite these challenges, these technologies, such as machine learning and natural language processing, have the potential to transform various aspects of the supply chain, including forecasting, inventory management, routing and scheduling, predictive maintenance and quality control. IoT has also enabled great visibility into the supply chain, providing ways to track and monitor goods in transit which helps manage expectations of actors in the end-to-end supply chain to the customer and improving transparency, trust and security.

The challenge for supply chain leaders is where to prioritize investments to gain and maintain a competitive advantage. They must find new ways to solve problems and balance what is asked of them from a sales perspective from commercial and marketing colleagues and from finance colleagues who are continuously challenging them to do more with less. The role of technology is no longer 'a nice to have' or a supporting tool in how supply chains operate, but is a key element when mixed with people, process and data to drive transformation. The winners will be those companies that can integrate emerging technologies with data to enable informed, data-driven decision-making, which optimizes the supply chain at a constant or reduced cost base, that can respond to changing market demands. If you don't embrace technology, you may find yourself behind the competition much quicker than ever before.

Endnotes

1 Byrnes, J L S and Wass, J S (2021) *Choose Your Customer: How to Compete Against the Digital Giants and Thrive*, McGraw Hill

2 Parke, M (2017) Thomas Friedman: Technology is accelerating faster than our ability to adapt. We can catch up, Working Nation, www.workingnation.com/thomas-friedman-technology-accelerating-faster-ability-adapt-can-catch (archived at https://perma.cc/X3A4-5W9Q)

3 European Commission (2022) Sanctions adopted following Russia's military aggression against Ukraine, European Commission, https://finance.ec.europa.eu/eu-and-world/sanctions-restrictive-measures/sanctions-adopted-following-russias-military-aggression-against-ukraine_en (archived at https://perma.cc/KHG5-APNF)

4 Ghaddar, A (2023) How Houthi attacks in the Red Sea impact shipping in the Suez Canal, Reuters, www.reuters.com/markets/commodities/how-are-red-sea-attacks-impacting-shipping-suez-canal-2023-12-18/ (archived at https://perma.cc/DRM7-YQPS)

5 Swanson, A (2025) Trump says he will announce reciprocal tariffs next week, *The New York Times*, www.nytimes.com/2025/02/07/business/economy/trump-tariffs-china-de-minimis-trade.html?searchResultPosition=16 (archived at https://perma.cc/Q49R-LSGY)

6 Gibbons, S (2022) How retailers should react to the decrease of in-store shopping, *Forbes*, www.forbes.com/sites/serenitygibbons/2022/03/03/how-retailers-should-react-to-the-decrease-of-in-store-shopping (archived at https://perma.cc/A79P-HB7C)

7 NRF (2024) NRF and Happy Returns Report: 2024 retail returns to total $890 billion, NRF, https://nrf.com/media-center/press-releases/nrf-and-happy-returns-report-2024-retail-returns-total-890-billion (archived at https://perma.cc/Z7NQ-CY8M)

8 CBRE (2023) The end of free returns: Will consumers continue to spend online? CBRE, www.cbre.co.uk/insights/local-response/the-end-of-free-returns-will-consumers-continue-to-spend-online (archived at https://perma.cc/4FGH-W4MN)

9 McGrath, M (2020) Climate change: Last decade confirmed as warmest on record, BBC, www.bbc.com/news/science-environment-51111176 (archived at https://perma.cc/RZ5Z-FANP)

10 Younger, S (2024) How crops and forests are responding to changing rainfall patterns, Stanford School of Sustainability, https://sustainability.stanford.edu/news/how-crops-and-forests-are-responding-changing-rainfall-patterns (archived at https://perma.cc/N7U8-Q97F)

11 Carey, H and Epifano, M (2024) The impact of natural disasters on economy and supply chain – and how to prepare for the worst, Thomas Publishing

Company, www.thomasnet.com/insights/how-natural-disasters-affect-the-supply-chain-and-how-to-prepare-for-the-worst (archived at https://perma.cc/N2NF-JB6F)

12 ITU (2024) Internet use continues to grow, but universality remains elusive, especially in low-income regions, ITU, www.itu.int/itu-d/reports/statistics/2024/11/10/ff24-internet-use (archived at https://perma.cc/TFZ3-7ZDN)

13 ITU (2023) More than three-quarters of the world's population own a mobile phone, ITU, www.itu.int/itu-d/reports/statistics/2023/10/10/ff23-mobile-phone-ownership (archived at https://perma.cc/UGY6-NGTK)

14 De Koster, R and Roy, D (2024) Research: Warehouse and logistics automation works better with human partners, *Harvard Business Review*, https://hbr.org/2024/06/research-warehouse-and-logistics-automation-works-better-with-human-partners (archived at https://perma.cc/C237-PYYH)

15 Rennie, E (2025) Gaining clarity: The power of supply chain visibility, Association for Supply Chain Management, www.ascm.org/ascm-insights/gaining-clarity-the-power-of-supply-chain-visibility (archived at https://perma.cc/59JH-9E9M)

16 Colehower, J (2023) Using technology to improve supply-chain resilience, *Harvard Business Review*, https://hbr.org/2023/09/using-technology-to-improve-supply-chain-resilience (archived at https://perma.cc/76AY-3NER)

17 Cason, P (2021) How the Internet of Things boosts supply-chain visibility, Supply Chain Brain, www.supplychainbrain.com/blogs/1-think-tank/post/32375-how-the-internet-of-things-boosts-supply-chain-visibility (archived at https://perma.cc/L5JL-LWZE)

18 Te Lindert, M (2022) Millions spent on supply chain technology fail to deliver, Supply Chain Movement, www.supplychainmovement.com/millions-spent-on-supply-chain-technology-fail-to-deliver (archived at https://perma.cc/T9YM-9FBY)

19 McCarthy, J, Minsky, M L, Rochester, N and Shannon, C E (1955) A proposal for the Dartmouth Summer Research Project on Artificial Intelligence, Dartmouth College, http://jmc.stanford.edu/articles/dartmouth/dartmouth.pdf (archived at https://perma.cc/9V3P-UEUS)

20 Google (2025) What is artificial intelligence (AI)?, Google, https://cloud.google.com/learn/what-is-artificial-intelligence (archived at https://perma.cc/B4AZ-SMYA)

21 IBM (2025) What is AI? IBM, www.ibm.com/think/topics/artificial-intelligence (archived at https://perma.cc/GUV6-MXS8)

The current state of supply chain management 03

Challenges and limitations

This chapter examines the current state of supply chain management, including common challenges and limitations, and the need for transformation.

CHAPTER AIMS

- Diagnose the common pain points and limitations of traditional supply chain management.
- Acknowledge the need for transformation and innovation to stay competitive.
- Uncover opportunities for improvement and growth in supply chain management.

'You can't be that kid standing at the top of the waterslide, overthinking it. You have to go down the chute.'

TINA FEY[1]

Over the first two chapters, we have looked at the evolution of supply chains and begun to position the role of technology within them as an element of competitive advantage. However, writing about it is easy; putting it into practice is sadly not. Supply chains enable companies to deliver goods and services to customers efficiently and effectively; however, approaches to managing supply chains have often been, and still are, found wanting by common pain points and limitations that hinder companies in achieving their full potential.

How 'traditional' supply chains have evolved

Traditional supply chains focus primarily on their role within the ecosystem from source to customer, often operating in 'firefighting' mode, responding to whatever is happening in the here and now. In the traditional model, the supply chain operated in a linear fashion and was executed by companies engaging with suppliers and customers on either side, but they were not so interested in other parties beyond. One of the reasons for this was visibility and transparency (key terms discussed in detail in Chapter 4) across the end-to-end supply chain where companies were unable to track inventory, shipments and orders with any level of confidence beyond daily views. Systems could provide a record of transaction; however, products could go missing between movements for days. Therefore, it was not odd to see staff walking around a warehouse and manufacturing areas trying to locate products. Transparency was at risk as well because at times the only evidence you had was of an order acknowledgement from a supplier and many emails or phone calls were required to gain assurance that what was being asked was being made and shipped to you.

Supplier performance was heavily biased on what was happening day-to-day. Their risks were unknown until something significant happened, which would usually lead to a significant escalation to a C-suite executive and more firefighting. Communication and collaboration were difficult, as manual ways of working were standard, with lots of written papers between parties. For example, a delivery note was written out and then uploaded onto the system, where the risk of errors and delays would occur. Keeping colleagues updated on issues was challenging, with meetings after meetings to resolve the issue of the day, leading to delays with everyday work. All of this meant companies struggled to respond to changing market conditions: supply chain structures and processes were designed for a different era and the ability to adapt to changes took time. This could result in having too much inventory, for example, resulting in higher costs and inefficient supply chains where products got to customers at any cost.

The building blocks of the supply chain have remained somewhat intact over the years because they have followed a linear traditional approach in which we were able to demarcate each subfunction around specific activities. The Supply Chain Operations Reference (SCOR) model in its original format, gave a solid standard foundation of Plan, Source, Make, Deliver, Return[2] with processes, activities and KPIs under each. It was originally developed in 1996 by PRTM (later acquired by PwC) and vetted by companies including IBM and Procter & Gamble. These aspects could be applicable to many different types of sectors and companies (with slight deviations needed for retail) but you could still find a path through using it. This has evolved to a more synchronous network view,[3] SCOR DS, which considers more elements aligned to the demands of supply chain professionals. We can still identify the key activities that are under the remit of supply chain with the additional elements of resilience, economics and sustainability. The updated processes include retail, omnichannel, strategic sourcing and supply chain strategy.[4]

We see challenges to outlawing traditional approaches to managing the supply usually where companies have been fixated on covering the cracks with plasters of capabilities rather than doing surgery. To move forward, companies need to take the lessons from the attempts of supply chain transformation and adopt an approach that is digestible and a vision that matures and advances to stay competitive. Companies must have a culture where transformation is part of the DNA, with endless curiosity to be outward-looking at ways to innovate and keep improving the supply chain. To address common pain points and limitations is to look at what capabilities you have and measure their performance. Then develop and adopt new technologies, processes and strategies to address them. Leading with technology is dangerous and can quickly result in a 'solution looking for a problem'. Therefore, taking time to assess and have a vision that stakeholders buy into is an important first step. At the same time, there needs to be some realism on what to go after; I have seen many companies that are good at unconstrained thinking and having lists of initiatives, however, when asked to explain the objective of them and how they individually or in conjunction help the supply chain achieve their strategy, the conversation becomes difficult. Roadmaps are good but in supply chain, we need to think about it as one large jigsaw: how each piece helps add to the solution and the sequence which gets us there with the freedom to reprioritize or add new pieces as we progress.

We have already touched on some of the themes we need to be thinking about in next-generation supply chain and the importance of technology to achieve them. However, for now, think of them as destinations rather than trying to provide specific technology solutions, because they each require the four elements of people, process, technology and data combined to achieve the desired outcomes. Below are the five themes that, by making conscious decisions in relation to them, will help your supply chain be competitive.

1. **End-to-end visibility:** Tracking and monitoring in near real time of inventory at its various stages from supply base through to customer.
2. **Collaborative communication:** Ability to always communicate via different forms (not just phone and email!) to relevant parties in the end-to-end supply chain.
3. **Agility and flexibility:** Able to respond to changing market conditions and customer needs quickly.
4. **Cost optimization:** A supply chain that optimizes execution to as low a cost point as possible, which is sustainable with understanding of trade-offs of not being efficient.
5. **Resilience:** Secure by design supply chain, redesigning them to proactively identify and mitigate security and risk concerns.

A suitable target for each must be decided because if you want to optimize cost in the supply chain, it could mean having to forsake resilience by reducing inventory to minimal levels or intentionally having a single-source supply and leveraging volume to try to negotiate for the lowest cost point. From a visibility perspective, how much is enough? Does your company need to know what a third-tier supplier is doing? Does that add value or a distraction? To achieve a service level of 95 per cent, this could mean having to invest additional cash by having more inventory in the system or having to add new assets such as a manufacturing plant or warehouses to achieve that goal. These points and many more need to be discussed and agreed upon.

These points can be examined through a decision tree (see Figure 3.1), which tries to find the balance between service, cash and cost, supporting what the company is willing to do to achieve competitive advantage.[5]

Service in terms of the supply chain can be defined as the fulfilment of customer orders aligned to agreed timelines. Companies are willing to go to whatever lengths to try to achieve this, which can be expediting orders

Figure 3.1 Supply chain triangle of trade-offs

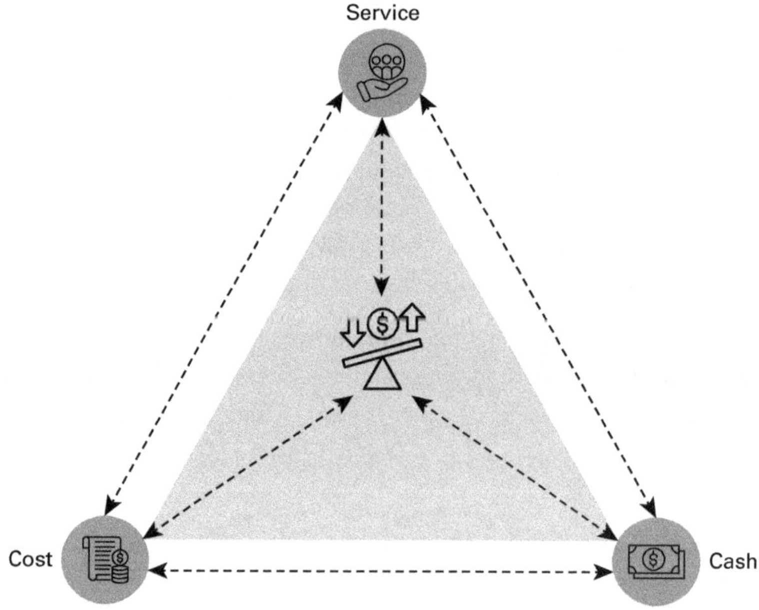

SOURCE Reproduced with permission of Kogan Page. DeSmet, B (2021) *The Strategy-Driven Supply Chain*

(e.g. using air instead of other slower speeds of shipping), un-reserving inventory for other customers and reallocating to important customers or changing production plans (impacting production efficiencies) to manufacture for this customer. Either way, this requires intentional interference in the supply chain to force an outcome.

Cash is a critical component of a company's supply chain, as it directly impacts the ability to manage inventory, pay suppliers and meet customer demand (aligned to the service promised). The importance of cash in supply chain decisions cannot be overstated as it affects the overall efficiency, effectiveness and sustainability of the supply chain. For example, when managing inventory, if companies don't have sufficient cash, they may be unable to purchase raw materials to make finished goods and therefore will be unable to fulfil customer orders leading to lost sales in the short term and the potential long-term loss of customers. Or, if they have too much inventory, then cash is tied up leading to potential increases in stockouts and lost revenues if they cannot purchase the inputs for finished goods desired by customers.

Cost of capital is the last part of the triangle and has a direct impact on the financial performance of the company since it represents the cost of obtaining

Figure 3.2 Supply chain pentagon of trade-offs

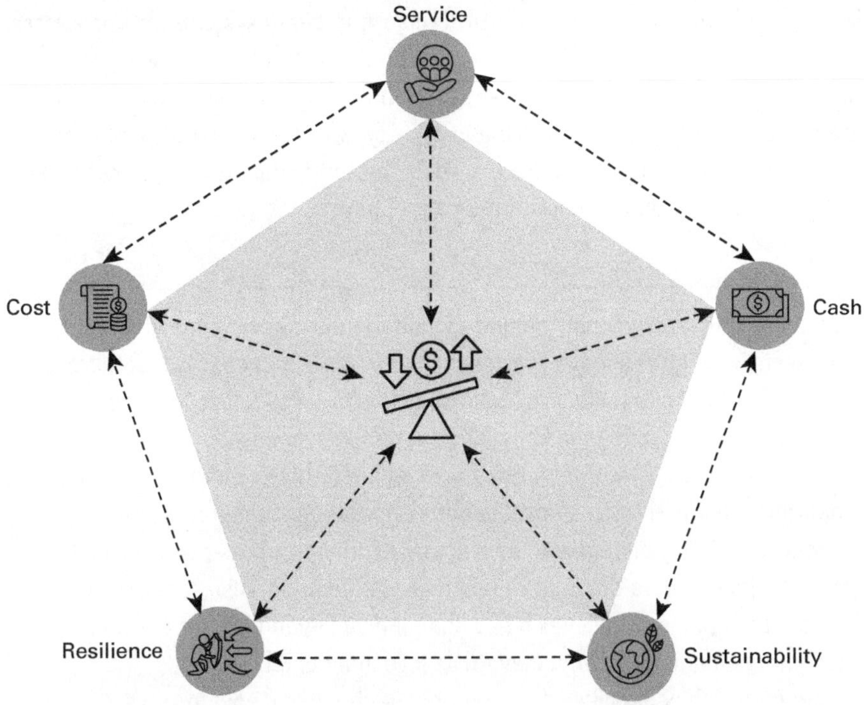

and using money to fund the finances of the supply chain operations. Capital allocation is one of the key decisions companies need to be making when looking at different supply chain initiatives. Leadership will be looking to prioritize investments that make higher returns in shorter time periods[6] and this is a factor in why technology investments require rigorous business cases in supply chains. Also, it is an opportunity cost, since doing one initiative might mean that the company cannot do another investment. This can be quite a subjective area when considering all the elements of the supply chain and lead to shelving the implementation of certain capabilities.

Where I deviate in thinking from the triangle is in the belief that the elements of resilience and sustainability need to be added, therefore making it a pentagon of factors that need to be balanced off for decisions (see Figure 3.2). Resilience in a supply chain context is being able to make decisions that help protect the company in the case of an unexpected disruption,[7] which could incur higher costs to serve customers. The need to have a supply chain that can demonstrate resilience has come into sharp focus since the Covid pandemic and has stuck during events since (e.g. conflicts and tariffs) where many organizations have struggled to cope.

Sustainability in the context of supply chain technology is the economic and environmental considerations that must be debated and agreed in the design, implementation and evolution of supply chain systems and processes. The overall focus of sustainable supply chain technology must be to minimize the impact supply chain operations has on the environment, economy and society, while still maximizing efficiency, effectiveness and profitability.[8] Companies do not have the luxury of being ignorant about the end-to-end supply chain, as I will discuss in detail in Chapter 8.

When we talk about supply chains as modern and leading edge, the reality (especially when you leave the dominant players in a sector) is that there is a mix of outdated and modern technology plastered together with the focus dominantly on cost and the knowledge in peoples' heads.

The opportunity for transformation in technology can be triggered via an operating model refresh, consolidation of operations or mergers and acquisitions. This change can be a catalyst for improved capabilities in the supply chain, where the focus again could be cost-led. Also, it depends on where the company believes it has competitive advantage. For example, take a food manufacturer: it may focus investments on R&D and manufacturing capabilities, being content with current cost to serve position from a logistics perspective.

The need to rethink supply chain

Technology within supply chains is messy as I have already stated, with today's supply chain architecture using 'plasters', bolting on capabilities, which has led to complex technology landscapes. Although it is difficult to be exact, Dimensional Research states that 9 out of 10 companies are unable to pull real time insights from ERP systems,[9] 82 per cent of companies are making decisions based on stale information[10] and almost 40 per cent of perceived data-driven companies have more than 50 distinct data silos they are managing.[11] A study with over 1,000 senior executives revealed that data-driven companies are three times more likely to report significant improvements.[12] What you will typically see, is a company that has some (if not all) of the following symptoms:

- Lots of disparate data sitting in many different systems.
- Data trapped in systems that are used for specific tasks only.

- Legacy systems of different instances with limited synchronization.
- Inability to have data that is shareable beyond the company and would rather provide in spreadsheet format.
- Variable latency (i.e. time for data to travel from one point to another isn't consistent).
- Data that is reactive and backwards-looking informing forward-looking decisions.
- Multiple versions of the truth.

For example, a well-known European Retailer I have worked with in the past cannot follow a product through their own part of the supply chain technology systems because they have the product code in one system and another code for the same product in a different system.

It can seem right at the time to enhance the supply chain with additional capabilities. However, if you do this without reference to the five themes stated previously in the chapter (i.e. end-to-end visibility et al), you could be adding an enhancement that does not bring value to the company and is just another application a user needs to open and look at daily. As Steve Jobs once said, 'I'm as proud of what we didn't do as I am what we did',[13] which is something supply chain leaders should be reflecting on when making decisions on capabilities.

If we take the five themes and try to connect them as a cohesive unit, we can start to think of them as a next-generation supply chain that aspires to being intuitive with the needs of the user, secure by design and customer-centric. Supply chains need to be simultaneously global and local, with the ability to adapt to market conditions and with an appreciation that a 'one size fits all' approach will not always be the answer. Putting a solid architectural and process foundation in place to then build capabilities on top is key (see Figure 3.3).

We should start with a solid base where key records of transactional data and supply chain-specific data exist. It is imperative that these records exist in a quality format and are readily available by surfacing them into a model that translates data from various systems into a standardized, transparent form, which can be integrated together. Advancements in setting up these foundations – from data lakes to generative AI layers – are already accessible. There is no right or wrong approach here, whichever the company feels best to be able to surface data should be taken. However, don't be distracted with fixating all money and time here: get the data to a state where it can be acted on.

Figure 3.3 Vision of technology architecture

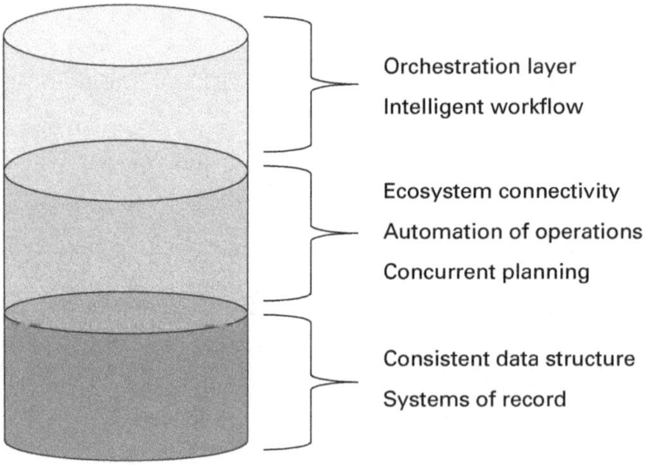

When the data is in one place readily available, then the base can be built on with three key elements of ecosystem connectivity, automation of operations and concurrent planning. In Figure 3.3, I have suggested these elements, but these as you can imagine, can be broken down much further and using the lens of what will provide most value to the company should guide priorities.

Planning is the avenue focused on connecting all data to inform and develop plans that all relevant decision makers agree to. Plans should feed other plans and be optimized to deliver what is demanded from the customer, primarily focusing on driving outcomes of value from service, cost and inventory perspective. Decisions regarding planning should be in relation to what levels of automation are appropriate, what interventions do we want our staff to be able to do and how will we continuously feed back into the system lessons from decisions made.

The next element is focused on the operations, covering the movement of goods in different states (e.g. raw materials, finished goods) and the transformation through manufacturing. Within this, companies must decide on levels of automation and the various hand-offs that take place across operations to reflect into systems such as an enterprise resourcing planning system where each transaction will be recorded and a warehouse management system, which will call and move materials to a manufacturing execution system (which consumes the parts that come from BoM to make a finished good), for example. The vision is to automate as much as possible to bring value through agility, flexibility and productivity. We will talk later about these and other systems in Chapter 4.

The last element looks at where the innovation of the cloud has changed how we view supply chains, prompting the move beyond siloed ways of thinking. The integration of an ecosystem to evolve the end-to-end supply chain with visibility and sharing of information to make improved overall decisions requires the sharing of data (internally and externally) so that we can engage at pace with other parties such as suppliers on changing requirements due to demand changes and/or manufacturing requirements, for example. This adds additional complexity as supply chain leaders must agree on guiderails about how different parties react to change, such as if they see requirements of over-agreed tolerance levels, they engage with customers directly rather than just act on the information. However, the goal is to improve the overall supply chain performance. The same applies downstream with logistics providers and customers, where in the case of customers, receiving data into the company from point-of-sale systems can improve the flow of goods to shelves, therefore reducing stockouts and maximizing opportunity for all parties.

These elements require decisions to be made about how to set up and run workflows. Often there is confusion when making a distinction between processes and workflows. The way to think about it is workflows are focused on step-by-step completion of a set of tasks that make up part of a process, while the process is to achieve a company goal (see Figure 3.4).

Most companies have process flows established. They typically sit in a large folder, which is sometimes dusted off and that is a problem. Process flows shouldn't be static – there will always be source to pay in 'purchase to pay'-type processes, for example. However, the path to achieve them should be challenged and refined continuously.

With processes having been established, they should be evaluated to understand which workflows within them could and should be automated to make them intelligent. This can be a significant unlock for companies, however it requires an understanding of the workflows and which data paths to

Figure 3.4 Processes that become a collection of workflows

prioritize – process owners should know this. Also, the norms of processes that span functional boundaries should be challenged because achieving automated workflows requires the buy-in from various functions in many instances. Companies should consider whether global standard processes are beneficial. There is certainly a need to document processes and share best practices, but companies must be careful: these processes may work well for a complex profitable business unit but might be over-engineered for a simpler business unit where the complexity does not exist.

On top of these intelligent workflows is an orchestration layer, a 'decision system' if you will, which provides visibility over the supply chain, assisting users with flagging issues / exceptions (such as inventory shortage) and guiding them to where they need to focus their time to make decisions. By having the visibility across all processes, users can make decisions understanding what is going on across the end-to-end supply chain rather than in isolation of others. We will discuss this in detail in Chapter 4 and will see in Chapter 13 through an IBM Infrastructure real-world example, how technology advancements can allow organizations to operate above systems, where we don't need colleagues to be experts in a specific system, but an expert in the supply chain processes and decision-making surrounding them.

This vision can be hard to imagine as you wrestle with the systems you have today; however, by looking at digital transformation through the lens of processes, getting clear on how each system and users interact with them, you can quickly start to understand what data is required and challenge whether the current data fulfils requirements. Then, by running the data through each system, it will become obvious whether there are constraints or an improvement needed. The mindset of the users and the embracing of the vision, understanding that their role will change will most likely be the main obstacle to overcome and we will spend time on that in Chapter 12.

We have a vision of a supply chain: we now want to aspire to it with technology. This aligns to the five themes we are focused on delivering to provide the company with competitive advantage – but how do we get from where we are today to that future, knowing that it will take time?

Transforming supply chain architecture

Baseline activities are table stakes when imagining a future supply chain with technology at the heart of it, as can be seen in Figure 3.5. Where we need to be careful is spending all our time trying to get these activities perfect, since at best, from a supply chain perspective, it might keep you surviving but not much more.

Figure 3.5 Pathway to next generation supply chain

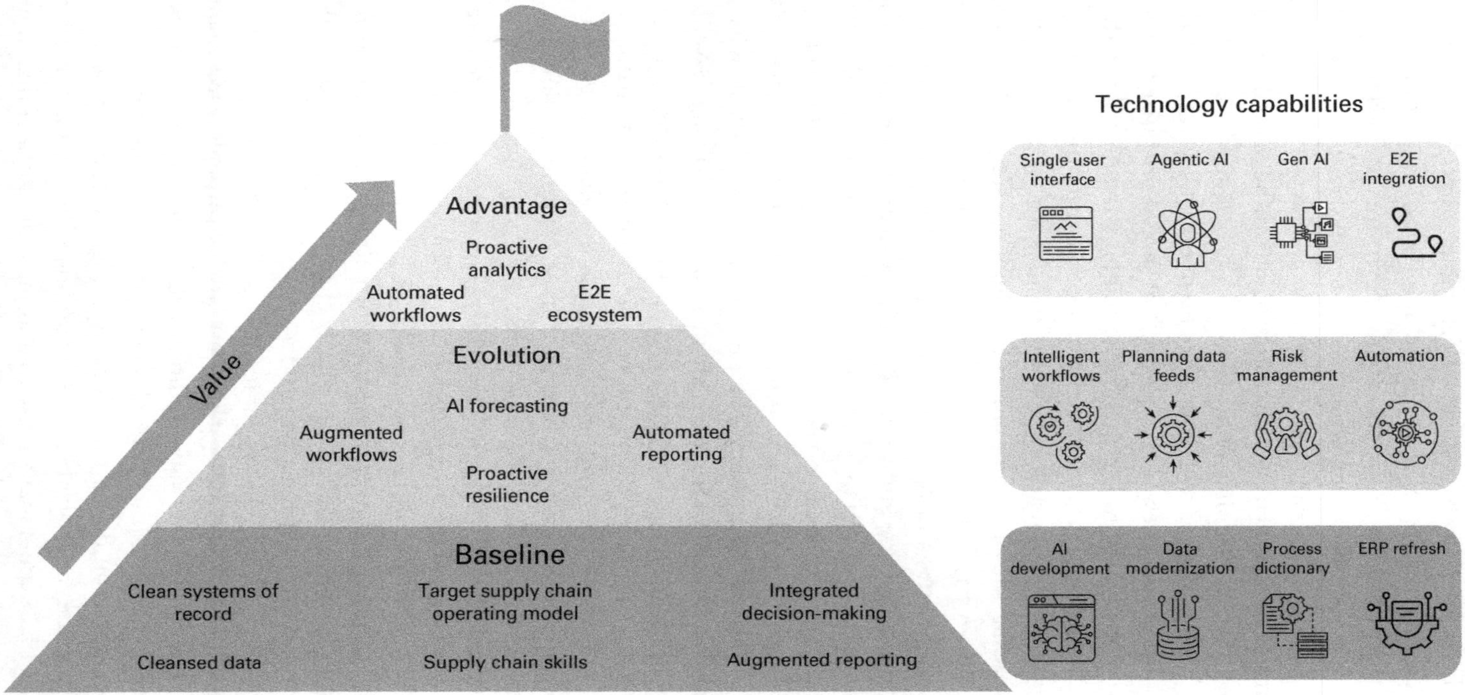

Data is the obvious first stage: to determine what you can surface via the different systems you have at play within your own control and what information you need from upstream and downstream as well. Focus on your company's data first. Determine the quality of data by reviewing and ranking it based on completeness to provide confidence to use it across other systems. This will help you focus on what can be moved forward now to create value versus other areas where work is required. Don't get fixated on getting everything perfect (a message I will repeat several times through this book) – process review and improvement should prioritize what needs to be done and to what level of quality. Seeing how the data flows through each system will help you understand whether opportunities are being missed to create value by not using specific capabilities within each system.

Assessing the operating model and how the supply chain is set up will help shine a light on weak spots (e.g. have you a strong balance between people, activities and value?) and where engagement is required with other functions to achieve end-to-end confidence in processes, as well as whether the model is appropriately set up to achieve desired outcomes. During this analysis, skills you have versus skills you require will begin to surface. Also, determining how decisions are made across processes will help inform levels of integration but more importantly, show what needs to change in future to improve them. Having a strong baseline will be achieved by a thorough review of the main systems that provide information to run the supply chain, process simplification and data.

At the next stage, you are looking at the elements that are going to accelerate you to a future state. As I have mentioned above in the vision, processes and finding the best approach with human and technology augmented to run workflows should be investigated and reimagined. Also from a system perspective, you need to understand whether all elements are covered within the architecture today or if there are duplicating activities occurring. During supply chain transformation, the point is to simplify the technology aligned to what the supply chain must do. If the technology isn't adding value, then it must be removed.

Review your company's data feeds – are they the appropriate feeds to help create value? For example, if we look at this through a forecasting lens, it might make sense to bring in additional data feeds (e.g. weather, local events) to complement the historical sales data you have available via the ERP, plus any analyses you do on this. Then you might reconsider how to complete these activities and whether they are adding value having the users involved in all parts. It could make sense to automate the steps of digesting data and doing the initial cut of the forecast rather than having a planner involved at that stage. This will then free up the user to focus more on value-adding activities. At this stage, it may also make sense to determine if a set of activities or even processes are best managed via outsourcing: these activities or processes are important but the current cost to deliver is prohibitive or the value is not material to the company. The underlining enablers of this stage are modern ERPs, workflows, automation and AI.

Like the move from having baseline to future state capabilities, it is not about mastering all at the future state stage before heading for advantage phase, when the company has its people, processes, technology and data working in sync. It uses technology and data to orchestrate across the end-to-end supply chain and is integrated with other parties. The company is forward-looking and able to operate with employees who know where to focus, based on what is happening in near real time within the supply chain. Use your judgement to decide when the company is ready to move to this stage.

It can sound far-fetched to think of this and the initiatives that can be achieved at this stage. The reality of ever-changing markets and customers will mean there will be elements that you may never get to a differentiating level, which may be acceptable to the company. However, remember that the goal should be to minimize effort on non-value-adding activities, finding ways to automate or, in the worst case, outsource. The ability to operate the supply chain through a single interface is where I can imagine a future and we will cover examples where companies are moving towards this later in this book.

Where next?

In this first part of the book, we have reviewed the technological innovations that have led us here to form the current supply chain architecture. I have provided a view of where supply chains need to go next from a technology perspective.

Traditional supply chains needed predictability and stability, which arguably is always going to be challenging to achieve. The traditional technology stack has had its time – yes there will always be a need for recording transactions, for example, but how the supply chain of the future needs to be set up must evolve. Having supply chain professionals opening more than nine applications daily to do their day job,[14] working off spreadsheets and manually doing activities was appropriate in the past, but with the advancements of technology, should we not expect the technology to do more of the heavy lifting of work for us? This is when we diverge between the systems of today to a future stack and in many instances, companies must be brave and set their own path.

The choices consumers have available will challenge brands to determine what the elements are that bring them success. Many will argue a strong brand alone does this and maybe does for companies such as Coke or Nike, but for the majority, they must have a proposition that delights consumers. It is crucial that the consumer gets their item when they want it and at a price point they are happy to pay, which is why the supply chain should never be underestimated in the DNA of a company. It was considered as a background function; an old colleague referred to it as the 'middle office' which I strongly reject. It is front office because if you fail the consumer on service, they don't forget, leading to reputational damage.

We have outlined a few considerations that affect modern supply chains and how they operate in a world that is increasingly complex. The role of markets and customers within this and how their expectations have changed over time (and will continue to) means that to achieve the nirvana of competitive advantage, companies need to have supply chains that can bend and bounce. This has been a struggle and many of the technologies provide parts of the jigsaw, but as you put them together, the picture is anything but clear. What has already been established is that technology has a critical role to play but it is not the solution alone. Its role within the dynamic of people, process, technology and data in unison will help to establish advantage. By thinking of them as a collective will help supply chain leaders build a stronger supply chain blueprint. The emergence of data-driven decision-making is only possible with technology and being able to provide near real time information to help users has been an exciting step.

It will be interesting to look back on this period and how much time was spent trying to surface data and that being deemed the breakthrough, instead of it being about what data with technology will help us make better decisions. Data needing to be taken out of the technology stack and worked on at the side sounds far too familiar to our old friend the spreadsheet, highlighting the need for technology to support data-driven endeavours.

Emerging technology is here under the grand title of AI. Everything today seems to have AI labelled to it, even systems that have been on the market before AI became popular in the supply chain, so no wonder there is plenty of confusion and scepticism. It does not help when solution vendors, the consulting community and research companies hype it up. However, AI and its different forms from machine learning, natural language processing, generative AI to agentic AI and beyond (all of which we will expand on later) have the potential to help supply chain professionals. From the desire to gain visibility in near real time (a phrase that will come up often in this book), helping predict what could happen and the 'so what' of those scenarios, to automation and optimization, will help companies become more resilient.

In the next part of this book, we will look at different technologies that companies are using or considering helping them evolve their supply chain today. We will explore visibility and transparency first, understanding the strides made to improve these and the pitfalls that come from hyping a technology which can end up following Amara's law[15] where we overestimate the effect the new technology will have in the short run and underestimate it in the long run. Then we will look at the role of data analytics with predictive and machine learning, where we will cover AI and the role it provides in helping advance information for supply chain professionals to make better data-driven decisions. Then we turn to the role of automation and robotics in supply chain. Digital platforms will be reviewed, which moves the supply chain beyond the traditional four walls, where collaboration and integration can exist, all driven by the innovation and adoption of cloud computing. Prior to cloud, these types of activities occurred but they were people and process-driven, with much confusion at play between discussions of different parties. As I remember, they could take over your entire day of work if not careful, such as the time when others and I were changing our minds by the hour on whether we needed a specific raw material! The cloud platform emergence has been a revolution and what it can enable is exciting for visibility and accelerating the way we can run supply chains. We will conclude this part with exploring how technology in the supply chain can positively impact companies' sustainability efforts and how they conduct their business in a socially responsible manner – these are table stakes and should have always been.

From there, Part 3 will look at *how* to go about implementing technology capabilities in supply chain. Firstly, taking apart a business case, we will then challenge how to measure success and what are the key metrics to look for improvements with. Then we will explore people aspects in relation to adoption, implementation and change management, because as exciting as implementing technology in supply chain is, there are many reasons projects fail: they need to be acknowledged so that you are armed with ways to approach it differently to increase your odds. Also, we will look at how to succeed in supply chain technology implementation, deep diving into various areas, (e.g. integration, configuration and data) to provide recommendations on how to do things differently. We will end that part looking to the future, what trends are emerging and what this means for supply chain professionals. We will bring to life the role innovation and growth should play within that in order to stay competitive.

We then conclude by bringing all the key elements from these sections together to cement the compelling reason why leveraging technology provides competitive advantage to companies and provide justifications for why innovation needs to be accelerated in supply chain.

Throughout the book, we will also stop to reflect on different companies and what they are doing with transforming their supply chains through the lens of technology with the purpose of these to bring to life the concepts and ideas since 'in theory there is no difference between theory and practice, but in practice there is' as overheard at a Computer Science conference,[16] and also widely attributed to Yogi Berra. Furthermore, insights from others who have lived and breathed supply chain and enabling them with technology are providing perspectives as well.

Where companies will need to adapt their thinking is on always wanting a proven case before innovating. However, to achieve a differentiated supply chain, we will need to take calculated risks. This is why being clear on your supply chain strategy, understanding how that links to the overarching mission of the company and having a technology landscape that links together and provides supply chain colleagues with the ability to spend their time working in one system is crucial.

Conclusion

Traditional supply chain technology and the role it plays is no longer sufficient to meet the demands of operating in today's complex and interconnected world. Companies must firstly do an outside-in view of their supply chain strategies to understand what is holding them back from being successful – which capabilities are causing them unnecessary problems or limiting them? Then through the multi-faceted view of people, process, technology and data, companies can determine how each of these separately and in tandem orchestrate ongoing success. A key part of that is leveraging innovation through technology so that data can be used to drive more of what we don't know today to aid the supply chain professional. Data has been proven to be in abundance (in varying levels of quality) but without technology, it is like turning a tap on and letting water pour out without a bucket.

Technology will be an enabler to help companies drive growth and competitiveness, allowing them to discover new ways to improve their supply chains so they are resilient, agile and sustainable, but as I will occasionally remind the reader quoting the Norwegian educator and Nobel Peace Prize winner Christian Lous Lange, 'technology is a useful servant but a dangerous master'.[17] We need to keep reminding ourselves technology alone is never the answer.

I imagine a future where supply chain professionals engage with a single user interface to operate the supply chain. Being an expert in a specific system solution will not be the role of a supply chain professional. It will be their ability to engage through a bespoke platform designed for your organization, asking questions to get the right data and with critical thinking, collaborating with colleagues and other parties, making informed decisions at speed.

Endnotes

1. Fey, T (2012) *Bossypants*, Sphere, London
2. White, S K (2021) What is SCOR? A model for improving supply chain management, CIO, www.cio.com/article/222381/what-is-scor-a-model-for-improving-supply-chain-management.html (archived at https://perma.cc/8T7N-CCD4)
3. Association for Supply Chain Management (2022) The SCOR digital standard (SCOR DS), Association for Supply Chain Management, www.ascm.org/corporate-solutions/standards-tools/scor-ds (archived at https://perma.cc/YBF3-3L8X)
4. Association for Supply Chain Management (2022) ASCM releases new SCOR digital standard, www.ascm.org/globalassets/ascm_website_assets/docs/9.19.2022-scor-press-release.pdf (archived at https://perma.cc/B7SZ-H4U3)
5. DeSmet, B (2021) *The Strategy-Driven Supply Chain: Integrating strategy, finance and supply chain for a competitive edge*, Kogan Page, London
6. Dutzler, H, Houck, B and Livingstone, J (2022) The smart moves your supply chain needs now, PwC, www.pwc.com/gx/en/issues/transformation/smart-moves-your-supply-chains-needs.html (archived at https://perma.cc/74KE-29N2)
7. Guo, Y, Liu, F, Song, J-S and Wang, S (2025) Supply chain resilience: A review from the inventory management perspective, *Fundamental Research*, 5 (2), www.sciencedirect.com/science/article/pii/S2667325824003108 (archived at https://perma.cc/96FB-VCZY)
8. Hariyani, D, Hariyani, P, Mishra, S and Sharma, K M (2024) A literature review on green supply chain management for sustainable sourcing and distribution, *Waste Management Bulletin*, 2 (4), www.sciencedirect.com/science/article/pii/S2949750724000993 (archived at https://perma.cc/AB5A-B3P8)

9 Fivetran (2022) Lack of real-time ERP data leads to business risk, poor decisions and lost revenue, Five Tran Dimension Research, https://cdn.prod.website-filevs.com/6130fa1501794e37c21867cf/665ddec5f9b64d134d4632fe_Fivetran-ERP-Report.pdf (archived at https://perma.cc/8N9R-2BJ9)

10 Perich, R (2022) Over 80 percent of companies rely on stale data for decision-making, Business Wire, www.businesswire.com/news/home/20220511005403/en/over-80-percent-of-companies-rely-on-stale-data-for-decision-making (archived at https://perma.cc/CMP6-KJFC)

11 Rafferty, T (2024) Harnessing AI, data and technology for smarter, sustainable supply chains, Tom Rafferty in conversation with Kayla Broussard, *Sustainable Supply Chain with Tom Rafferty*, [podcast], 16 December 2024

12 Farrell, M (2023) Data and intuition: Good decisions need both, Harvard Business Publishing, www.harvardbusiness.org/data-and-intuition-good-decisions-need-both/#_edn1 (archived at https://perma.cc/RPG3-UYYH)

13 Cho, M (2016) How Steve Jobs saved Nike (and Apple) with 1 simple piece of advice, Inc., www.inc.com/mikael-cho/how-steve-jobs-saved-nike-and-apple-with-one-simple-piece-of-advice.html (archived at https://perma.cc/4HYQ-LGXS)

14 Darrow, B (2017) App fatigue on the job is becoming a problem, survey says, *Fortune*, https://fortune.com/2017/09/14/business-pros-app-fatique (archived at https://perma.cc/9Y35-76ZV)

15 Ratcliffe, S (2016) Roy Amara 1925–2007 America Futurologist, *Oxford Essential Quotations*, Volume 1, 4th edition, Oxford University Press, Oxford www.oxfordreference.com/display/10.1093/acref/9780191826719.001.0001/acref-9780191826719 (archived at https://perma.cc/63RE-XVHK)

16 Savitch, W (1994) *Pascal: An Introduction to the Art and Science of Programming*, Addison Wesley, Reading, MA

17 Lange, C (1921) *Internationalism* [lecture], The Nobel Prize, www.nobelprize.org/prizes/peace/1921/lange/lecture (archived at https://perma.cc/V49R-MYNB)

PART TWO
Digital technologies for supply chain advantage

Digital technologies for supply chain visibility and transparency

04

In this chapter, we will look at what visibility and transparency mean in supply chain terms and explore a variety of technology solutions that have emerged to address these needs.

CHAPTER AIMS

- Realize the importance of end-to-end visibility and transparency.
- See how digital technologies achieve real time visibility and transparency.
- Recognize opportunities for implementing digital technologies.

'Amateurs talk about tactics, but professionals study logistics.'

GENERAL ROBERT BARROW, USMC[1]

REAL-WORLD EXAMPLE The US Army – one of the most complex supply chains in the world

The elements of a complex supply chain include the need for global coverage, the orchestration of many different products and services, and the ability to be always ready to respond. The US Army is a key example of a complex supply chain, with a presence in at least 80 countries with 750 military bases (the exact numbers are difficult to determine and some information is classified).[2] Its supply chain is a critical component of the organization, responsible for providing equipment, supplies and services to support a variety of operations whether for combat or humanitarian purposes via a global network of bases and depots across a variety of time zones.

The ability to predict demand and plan accordingly is difficult due to the operations being inherently uncertain and volatile. Risks from cyberattacks are ever constant, with bad actors intentionally trying to hack systems, cause data breaches and disrupt moving products. Like any organization, it is also constantly being asked to justify budgets, which can reduce its ability to procure what it requires and keep up to date with latest equipment. Finally, with such a supply chain, balancing operational needs with environmental concerns can be difficult, with attention drawn to how it can reduce waste, energy usage and its carbon footprint.

The main components that make up the army's supply chain are the same as many other organizations, including demand planning (forecasting and predicting future requirements), inventory management (distributing the stock and optimal replenishment flow between the warehouses and locations where required such as in theatre) and supply planning (with the most notable difference in manufacturing being completed by a few specialized defence contractors).

The evolution of forecasting methods that better represent the reality of the non-linear supply chain (which tries to predict future demand without historical data to rely on as a starting point) that look at interactions, delays and feedback effects,[3] has had many breakthroughs. This has improved confidence in the forecast by using techniques that can better understand cause and effect. One model, which is described as 'generalised autoregressive conditional heteroskedastic' (GARCH) was tested and provided evidence of improved predictive performance during military operations in Iraq.[4]

The procurement of goods and services from the supply base can be complex. Contracts with private companies, agencies or international partners (i.e. other countries' armies) can be negotiated for significant long durations. For example, helicopters and tanks are not make-to-stock (MTS) items; therefore, each can have long purchase-to-delivery time cycles.[5]

From an inventory perspective, tracking and managing storage, distribution and maintenance of equipment and supplies are key activities. Knowing at any time what inventory you have and being able to confidently track it as it moves around the supply chain requires visibility and transparency because it can literally be life or death if this is wrong. For example, during Operation Desert Storm in 1991, planners established 60 days of inventory before commencing activities. This compares to the subsequent invasion of Iraq where seven days of inventory was deemed appropriate due to improved understanding of demand, predictability and stability.[6] This was due somewhat to the confidence of visibility in the supply chain.

Transportation feeds on visibility as well, since the army uses a range of different modes to move products across its supply chain and coordinate it all within appropriate regulations. Maintenance and repair considerations make the supply chain complex too, since keeping equipment in use requires routine checks, repairing damaged equipment and managing spare parts so they can be deployed at speed where needed, which in many instances can be in combat theatre.

Different technologies have already had a positive impact on the US Army's supply chain. The use of radio-frequency identification (RFID) tags, which hold encoded data and share this with other technology systems (which we will explore in more detail in this chapter), was a game-changer for the military. Then, take the establishment of the transponder system, which was a form of radar to identify friend or foe in the 1950s.[7] These systems are now able to understand whether products in shipment have any integrity issues (e.g. exposed to extreme temperatures in transit). Also, there are smart lockers that store and distribute items securely. Using sensors, they inform users of where their contents are and secures them so only a specific individual has access. Some examples of their contents include uniforms, personal kit and sensitive items such as laptops / phones.[8] The use of visibility here has helped reduce the impact of stockouts and helped evolve the process of replenishing inventory once they are taken from the locker for use.

When it comes to orchestration, the US Army has a control tower (a system providing real time visibility, monitoring and management of the supply chain). The details of it are unknown to us civilians, but we do know that it is helping it obtain a single source of truth throughout the supply chain and is improving its operational readiness by being able to integrate with a variety of systems to see assets, inventory and logistics in motion in near real time. One Network Enterprises, a vendor that has been involved in one of the US Army's programmes, enabled a 10–40 per cent reduction in inventory and manual planning reductions of 30–50 per cent with the use of its technology.[9]

What we can learn from the US Army is that its supply chain is critical to its success and being able to have visibility across its operations is table stakes. The ability to gain near real time information is a challenge so it has invested in technology and partners to help it. Further technological solutions it implements will continue to prioritize visibility, and the use of innovations, such as drones, suggests the desire for more autonomous and digitally enabled solutions to help its supply chain be even more responsive in its approach to modern warfare.

Visibility and transparency

Visibility provides all parties with a clear understanding of what is happening at any moment in time. This informs decisions and holds others to account, assisting with:

- Building confidence and creditability.
- Encouraging collaboration and participation.
- Promoting a culture of openness and honesty.

Visibility in supply chains can be focused within the four walls of the organization or can expand to 'end to end', which means seeing far upstream through suppliers' supply chains, as well as downstream moving through intermediaries (where applicable) to the end customer as shown in Figure 4.1.

Whatever the scope of visibility, the objective is to have the ability to see and understand supply chain processes from the perspective of physical flow of goods, information and finances. While this sounds simple as a concept, many organizations struggle to have visibility beyond immediate direct suppliers. In 2022, the McKinsey Global Supply Chain Leader Survey showed that only 35 per cent of respondents had visibility beyond immediate direct suppliers, but this figure had improved to 60 per cent by 2024.[10] This emphasizes the importance of having a technology stack that can orchestrate across the end-to-end supply chain to maintain a competitive advantage.

There are various ways to show different levels of progress when it comes to visibility. I find that when engaging companies on strategy projects, it is best to quickly determine the levels of maturity of visibility today (e.g. they know their suppliers, where their manufacturing is done, when orders are en route etc). It is also key to determine their aspirations because demanding end-to-end visibility might not be worth the cost it would take to achieve this.

Since the original supply chains were developed, we have focused on reacting and anticipating stages of visibility, as shown in Figure 4.2, stage 1,

Figure 4.1 Visibility in the supply chain

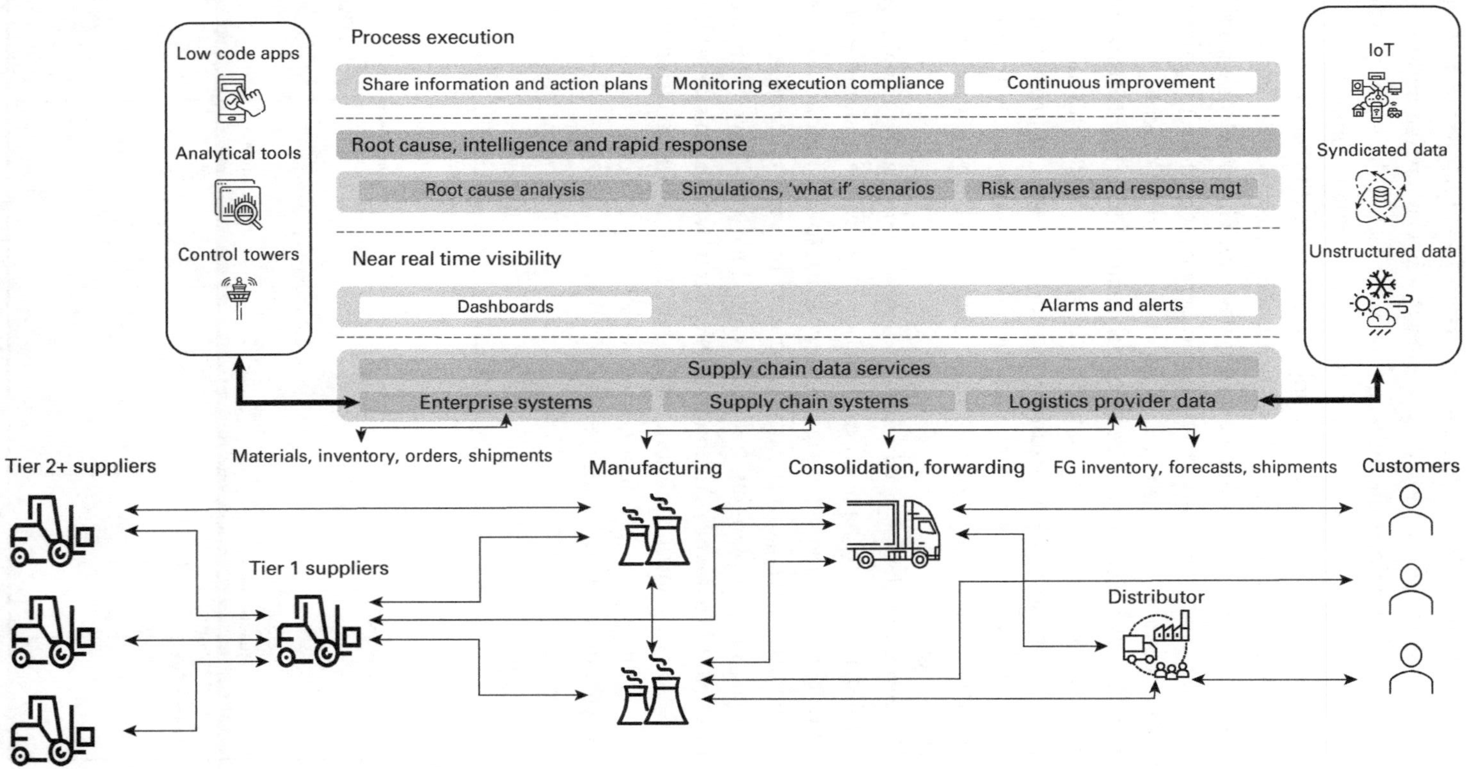

spending countless hours trying to understand what happened and why. The effort to collate information and present it was challenging and by the time parties had determined the answer, they were facing the same questions on another problem. These issues have been tackled somewhat with the introduction and evolution of dashboards, where different data is stitched together to give us views of the past as we move to stage 2. Alarms and alerts can be set so when something occurs, it flags up. The problem with this level of visibility is that you are reactive and dependent on keeping dashboards and data refreshed; otherwise, they become redundant quickly.

System advancements have made it easier for organizations to push forward into scenario management, which tries to predict (stage 2 to 3 in Figure 4.2) what is likely to happen and then, with improved analytics, answer questions about what you should do. Some analytics then go a step further, basing their recommendations on what you have done previously (stage 4).

However, the challenge when using these systems is 'explainability'. The user must be able to have an explanation of why the system derived that specific recommendation. Stage 5 moves us to an autonomous system. Autonomous systems are already being used by a few companies in specific situations, such as planning for specific products that have similar characteristics (stable demand). Although I believe it may take considerable time to get mass adoption of autonomous systems, it is worth considering if the supply chain will be predictable and stable long enough for this to be an option for most organizations.

Figure 4.2 Phases of visibility

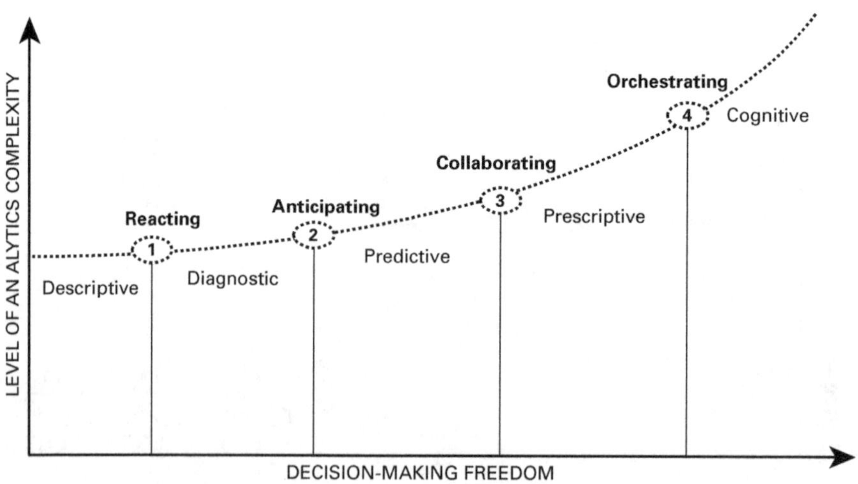

Transparency dives deeper than visibility by providing insight into the processes, decisions and outcomes leading to trust and credibility building with stakeholders. By adhering to this, parties are encouraging accountability and responsibility coupled with learning and a continuous improvement outlook.

The term transparency refers to being able to be transparent at each step of the supply chain, through the ability to trace and explain which parties were involved and what transformations occurred throughout the process of taking raw materials from their source through to a final product being in the hands of the end customer. In its simplistic form, this doesn't sound difficult; however, if you take, for example, a flavoured crisp, you can be operating a supply chain with excess of 100 ingredients, which when combined, make the product – then consider the packaging elements! This flags up other challenges inherent to supply chain such as resilience and how to cope with many actors in the supply chain, which we will touch on throughout the rest of the book.

When visibility and transparency are aligned, trust will inevitably be established. Trust is developed when companies demonstrate a commitment to openness, honesty and accountability, which in turn cultivates a positive and productive environment, where stakeholders have confidence and empowerment to contribute and make decisions.

In supply chains, the benefits of having visibility and transparency to support practitioners include the following:

Decision-making

With visibility, organizations can make informed decisions based on near real time data. Without this, decision makers are relying on assumptions or historical data, which takes time to assess and can lead to decisions being made that are out of date with changing conditions. When you consider visibility, the ability to deploy strategies for elements, such as demand forecasting, inventory management and logistics, will be improved.

Visibility and transparency enhance the operation and provide the ability to be strategic about how you want to evolve the assets in your supply chain to adapt to changing demand and risks, which will protect the business.

Speed to market

Having a supply chain where all movements are visible helps to identify bottlenecks and inefficiencies, which, when rectified, will lead to a more efficient supply chain. This will help reduce time to market and increase flexibility to respond to changing demand patterns.

Risk identification and mitigation

Supply chains inherently are susceptible to risk. Many factors affect this, from exogenous shocks to political instability to economic fluctuations. Take, for example, the risks of shipping products through trading routes like the Suez Canal, which has been in the headlines for several reasons since 2020, including 'Ever Given' vessel, which blocked the route in March 2021, holding up more than $5 billion in trade daily for the six days it was stuck[11] to Houthi rebels attacks, which started in winter 2023 and cost companies significant money and time (~10 days longer) to re-route.[12]

Changes in governments and regimes are challenging supply chain leaders and bringing back focus on strategies around dual sourcing (using two or more suppliers for raw materials) to nearshoring (bringing back manufacturing or another function to a country nearer to the home country) as examples.

By having visibility, organizations can proactively identify potential disruptions and implement a variety of contingencies to manage these.

Customer satisfaction

When companies have the confidence to see through their supply chain, it enables them to provide up-to-date, accurate information to customers, from availability to shipping times to order status. This helps enhance trust and loyalty in customers because it removes uncertainty in the order fulfilment process and can lead to a reduction in lost sales.

Data access

When companies have visibility, they gain unfettered access to a vast array of data. Valuable information enables the development of highly tailored insights, allowing businesses to make better decisions and increase their productivity with time spent on proactive activities.

Compliance

Regulatory compliance, while it can be different in approach across sectors, has been an instrument of change for how businesses operate. Having visibility and transparency helps companies maintain accurate records and traceability as well as meet regulatory requirements and demonstrate accountability.

Negotiation

Having information available and recorded in near real time helps procurement personnel understand how well suppliers are performing and gives them supporting evidence when negotiating product costs, placement and timing.

Inventory

One of the ways to compensate for a lack of visibility and transparency is to have higher levels of inventory (increasing stock levels to absorb and negate variability). If you can have these elements in place, then you will be able to measure lead times, enabling the company to constantly monitor and update lead times to reflect reality and not have to leverage inventory so much.

Digital technologies enabling visibility and transparency

The vast number of different solutions available to assist companies in establishing near real time visibility and transparency demonstrates how important these elements are to running the next generation of supply chains. Various challenges, which will be discussed throughout this book, include the need to have a consistent data approach to utilize from a smorgasbord of applications that exist within the supply chain architecture of most organizations.

Cloud computing

Cloud computing refers to the delivery of computing services and encompasses storage, processing power, software and networking over the internet (i.e. cloud). Prior to the cloud's existence, companies relied on local servers or physical infrastructure. Therefore, anything they wanted to do was restricted by accessibility, scalability and cost. I saw this first hand in my supply chain role at the beginning of my career, when we used paper templates, Excel and an ERP system to do every task in the supply chain.

In relation to supply chains, cloud computing enables companies to do many tasks: being able to gather, process, analyze and store more volumes of data from various sources. This means that demand forecasting, inventory management to supplier / customer collaboration and near real time visibility of products in transit is improved. Therefore, the cloud allows businesses to see their entire supply chain more clearly, helping supply chain managers make better, data-informed decisions, react faster to problems and optimize processes like order fulfilment, re-routing products and restocking inventory.

Supply chains are constantly shifting due to changes in demand, seasonality and world events. The cloud provides scalability, meaning companies can swiftly scale up or down based on their needs. This flexibility is especially vital for businesses with spikes in demand or fluctuating production timetables. The cloud also enables companies to rapidly integrate new supply chain partners, expand into new markets or adjust their strategies without major upfront investments in IT infrastructure.

One of the main benefits of cloud computing in supply chain is its ability to enable better collaboration and communication between stakeholders (internally via other functions or externally with third parties). In traditional supply chain models, data silos often exist between suppliers, manufacturers, distributors and retailers. Cloud platforms provide a centralized hub for all parties to access and share near real time information. This enhances co-ordination and helps reduce mistakes, delays and miscommunications. For instance, suppliers and manufacturers can easily discuss order status, delivery schedules and any potential disruptions.

Although security worries have traditionally prevented cloud adoption, modern cloud providers implement robust security measures to protect sensitive supply chain data. These include encryption, multifactor authentication, data backup and regular security audits. Many companies find cloud-based systems more secure and capable as part of an ecosystem than on-site solutions, which are often vulnerable to hardware failures and human mistakes. Additionally, cloud providers invest heavily in disaster recovery and business continuity to keep supply chain data safe and accessible even if systems fail.

Cloud computing's impact on supply chain has been transformational and it can be observed simply by the number of different providers of solutions in the market. Having only six main providers of ERP software on premise in 2000, the number has grown exponentially with 10 companies owning more than 70 per cent market share as at the end of 2024[13] (there are more than 250 vendors in the market).

Cloud computing does present challenges in the supply chain; in many instances, this is similar for all company functions when integrating cloud technologies. Legacy systems, which many companies have still at the core of their business, make integration significantly difficult from time, effort and cost perspectives. The perception of data privacy and security do still concern companies and while it is estimated by Parametrix that 93.6 per cent of the Fortune 500 use cloud technology, there are still some who are resisting.[14] Cloud computing needs internet connectivity; therefore, any poor or unreliable access can disrupt supply chains. This is prevalent where companies operate in remote locations with limited infrastructure.

Cloud computing has fundamentally transformed supply chain management by offering scalable, flexible and cost-effective solutions for optimizing processes across the entire supply chain. With enhanced visibility, near real time data access and improved collaboration, cloud technologies enable businesses to respond more quickly to market changes, optimize inventory and reduce costs. While there are some challenges related to security, integration and connectivity, the benefits far outweigh the drawbacks for most businesses. As cloud technologies continue to evolve, they will likely play an even more prominent role in shaping the future of global supply chains.

Internet of Things (IoT)

IoT is the collective network of connected devices and the technology that facilitates communication between devices, the cloud and the devices themselves. They include sensors and RFID tags that can be attached to products, shelves in stores and warehouses, pallets and transportation to provide monitoring information important to users in near real time and enable proactive communication with customers.

For example, when used on shelves, IoT devices can track stock levels and automatically reorder products when they run low. When products are being shipped, their location and the conditions (e.g. temperature and the integrity of products) can be monitored. A leader in the use of IoT is Walmart. It has over 7 million unique data points managed via sensors in its US operations that send more than 1.5 billion messages daily. These messages include information in relation to temperature, performance and energy consumption. Any issues raised based on these messages can alert the relevant parties to engage immediately. In practice, consider a freezer in a store: Walmart staff at any time can check if it is at the appropriate temperature and then, if there is an issue, have in place a process to quickly investigate and action a recovery.[15]

Caterpillar has been using IoT devices on its equipment to monitor an array of capabilities from location to fuel efficiency to improve maintenance, so it reduces downtime of equipment. By collecting vast data on its equipment, it can provide a platform for its customers to track and optimize use.[16]

The rise of online food deliveries took off during Covid[17] and is an excellent demonstration of IoT in action on several fronts, working from receiving orders through to fulfilment, building trust with customers who could track progress on their IoT devices throughout the process.

The IoT process, as shown in Figure 4.3, has many benefits:

1 **Ordering and reordering with IoT devices:** for example, smart speak or voice-enabled assistants can be integrated with food ordering applications, as well as personalized recommendations based on prior orders, preferences or dietary restrictions.
2 **Delivery tracking through using global positioning system (GPS) trackers:** this provides customers with the estimated time of arrival throughout the process and live tracking of delivery vehicles. The delivery driver can use route optimization software on an IoT device to choose an optimal route based on live traffic information.
3 **Temperature monitoring and quality control:** IoT sensors can monitor temperature and quality of food items (primarily perishable) during storage and transportation. This builds confidence and trust that food items are managed and aligned to food safety standards with any deviations triggering alerts to resolve any issues.
4 **Inventory management:** IoT devices integrated with inventory management systems provide live food item stock levels, expiration dates and re-ordering points so that the food company minimizes waste and replenishes ahead of demand.

However, the challenges faced with using IoT devices include the cost to maintain them, since the majority will at some point run out of battery, they can be costly to replace and they sometimes can be unreliable. Also, they can collect and store data, which may not be mission-critical, which can lead to increases in storage costs and raises privacy and security concerns.

Control tower

'Control tower' as a phrase can conjure a variety of meanings, but in the context of visibility and transparency, it refers to a centralized 'nerve centre' that provides near real time visibility, monitoring and management of the

Figure 4.3 IoT process

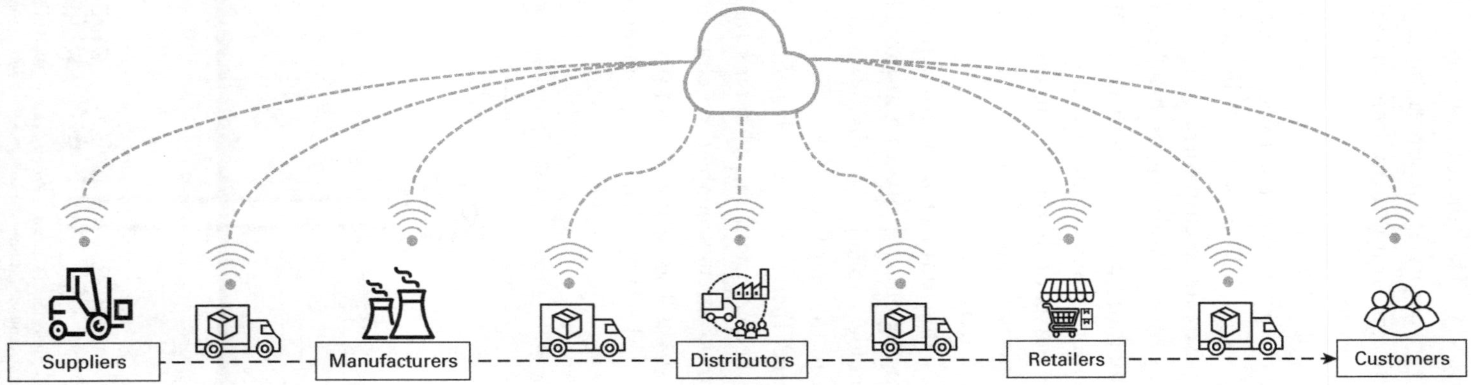

supply chain (see Figure 4.4). The control tower consolidates data from a variety of sources (e.g. inventory levels, order fulfilment, shipment status and demand forecast) to support informed decision-making and where possible, to optimize the operation.

How much of the supply chain is in scope is dependent on the decision maker, who can choose to keep the control tower specific to their part of the supply chain or go beyond their 'four walls' to be inclusive of upstream to supply base and/or downstream to customers.

Early versions of the control tower have been commercialized with a particular focus on using it to manage the movement of products with data feeds from a transportation management system (TMS) and logistics partners, as discussed at the outset of this chapter in the US Army example. The focus has been to inform interested parties of where inventory is and provide estimates of time of arrival, for example, in relation to cargo. Modern versions include additional features such as providing users with the ability to use scenario planning, conduct analytics and provide conflict rooms to bring interested stakeholders together to discuss and agree the best way forward for the situations they face.

As you would imagine, there are several benefits to implementing control towers. Firstly, in keeping with this chapter's theme, control towers provide visibility and transparency, enabled by cloud and IoT. Companies can track their supply chain in near real time, provide visibility into pertinent information to run the supply chain (e.g. inventory, shipments and

Figure 4.4 Control tower

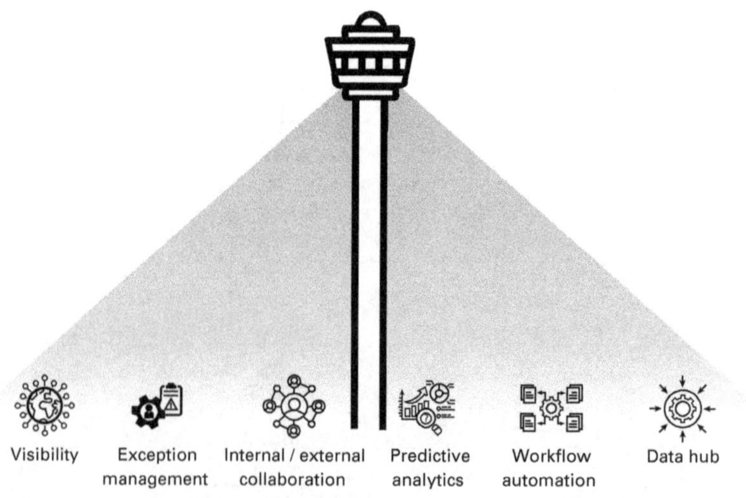

order status). Transparency is provided from being able to see where goods are at all times, as well as spotting potential disruptions and delays, so that the company can act on them proactively, re-routing shipments, rebalancing inventory across the network and turning on alternative supply to avoid stockouts. These powers give the supply chain more agility to bend and bounce to remain competitive. From a cost perspective, being able to optimize inventory and capacity in the network supports companies in identifying inefficiencies and quickly taking corrective action to minimize impact and reduce the cost to serve.

Beyond your company's four walls, engagement with suppliers will become more collaborative as information sharing will drive decisions in the wider supply chain and will lead to working together to resolve issues in shorter time horizons with less subjectivity. At the other end of the supply chain, customer service will improve as messaging to them will be consistent with what is happening, providing accurate tracking information and managing their expectations, leading to increased loyalty.

With any technology implementation, there are challenges with control towers, which should sound familiar already. This tool requires data integration and harmonization of multiple sources of data, which can be difficult and time-consuming to embed. Data security and privacy must involve encryption and managing users' access to safeguard sensitive information. The cost to implement the tool requires investments in technology, infrastructure and personnel to set up and run, which can be prohibitive in the short term to many companies. The tool also needs to be scalable and continue to evolve with business needs, which require ongoing investment. Finally, in my experience, the control tower as a concept has been heavily focused on the technology itself, with not so much consideration given to what people want to use it for, what processes it will cover and what data will aid the users. Therefore, it becomes nothing more than a visualization tool until all four elements are considered in concert.

Digital twin

Developed originally for manufacturing and the aerospace sector, this technology has become more popular in the overarching supply chain. Through simulating systems and operations from a virtual perspective, this tool enables businesses to gain insights in how their supply chain operates to improve decision-making and optimize them. Defined as 'a virtual representation or digital counterpart of a physical object, system or process'[18] it provides a replica that can be continuously refreshed from IoT devices and other data sources, providing users with the opportunity to simulate upside business potential or supply chain disruptions without interfering with them. It can model everything from a new product, to expanding the supply chain network and/or customer base, including nodes and movements between nodes. See Figure 4.5.

From a visibility perspective, digital twins can provide a view of a company or end-to-end supply chain by offering live insights into the flow of goods, transportation routes, inventory levels and integrity of shipments. By creating a digital replica of the supply chain, companies can track each product's movement, whether it is raw materials from suppliers, work-in-progress items in manufacturing or finished goods that are customer-bound. For example, IoT devices embedded in shipping containers or products can provide near real time data to the tool, updating its model with the latest information about location, temperature, humidity or potential delays. This visibility assists users to make informed decisions quickly, respond to disruptions and help them run their operations. From an asset perspective, it can provide predictive capabilities to perform preventive maintenance to reduce costly downtime.

Beyond replicating a supply chain, it can be a powerful tool by providing simulation capabilities, such as to test scenarios and identify and recommend routing, schedules and load configurations. For example, it could model traffic patterns and trailer capacity, enabling users to optimize delivery schedules live. If delays occur or new constraints arise (such as road closures or accidents), it can adjust logistics plans accordingly, helping to minimize disruption and reduce operational costs. Furthermore, it allows

Figure 4.5 Digital twin

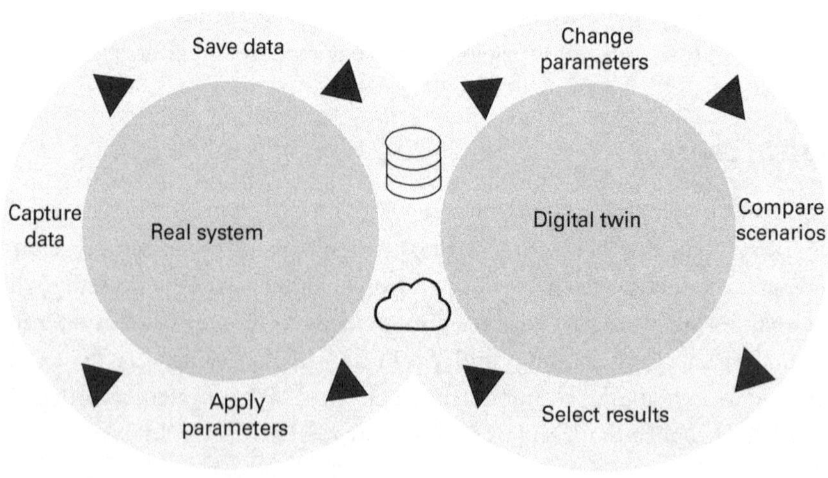

companies to evaluate alternative transportation strategies, such as switching to different modes of transportation (e.g. from trailers to rail to air), based on live data.

From an inventory and warehouse management perspective (as well as aiding in the creation of alternative warehouse layout plans) it can be significantly enhanced with digital twins. By creating a virtual representation of a warehouse, companies can simulate inventory flows, storage capacities and order fulfilment processes. For example, it can monitor the movements of products in a warehouse and automatically adjust inventory levels as products are added or removed. Near real time data on stock levels, demand forecasts and order statuses allows warehouse managers to optimize storage, reduce lead times and improve overall inventory accuracy. The model can also highlight inefficiencies, such as bottlenecks in picking and packing, and suggest improvements.

Further uses could be in relation to risk scenarios, where digital twins can model factory shutdowns or supply shortages, for example, evaluating how these events could impact the supply chain from a time and cost perspective. By running simulations under various conditions, companies can identify potential vulnerabilities and develop contingency plans to minimize disruptions. Working with suppliers and providing them access to the model, they could monitor inventory levels and delivery schedules, which allows them to better align their production schedules with demand. Similarly, third-party logistics providers can use the model to track shipments and optimize their transportation operations.

REAL-WORLD EXAMPLE British Sugar

British Sugar (a subsidiary of Associated British Foods PLC) is the sole processor of the UK's sugar beet crop. It was struggling with having to make many decisions manually, relying heavily on expert knowledge and intuition. Unplanned downtime impacted production, as well as concerns of losing technical knowledge due to employees' retirement. It wanted to address these challenges by developing a digital twin for its supply chain lifecycle, implementing predictive maintenance strategies and integrating operational data with AI. By doing this, it was able to simulate scenarios and optimize production without disturbing operations. This resulted in it being able to reduce failures and downtime from rotating equipment. The specific benefits included reducing decision-making time by 20 per cent due to having near real time data insights and reducing time for new employees to be become expert sugar makers by 50 per cent.

With the introduction of extensive modelling, it mitigated the risk of technical knowledge loss and halved the time it takes new employees to become expert sugar makers, enabling a smoother transition of knowledge, reducing the burden on experienced decision makers and preserving critical technical knowledge for the future.[19]

While digital twins are becoming more popular, they come with challenges including significant investment costs in putting together the infrastructure (e.g. IoT devices) and data storage and analytics capabilities. Also, the modelling element in relation to simulating scenarios will make the process time-consuming and require significant resources at the outset. However, they do provide a robust solution for visibility to virtually replicate supply chain processes and systems, helping companies monitor performance in near real time, supporting improved decision-making and proactively managing disruptions.

One of the questions I get asked frequently is in relation to what the difference between a control tower and a digital twin is. Personally, I feel the lines are often blurred between the two solutions, so the main differences have been outlined in Table 4.1.

I expect in time that the term 'control tower' will disappear in so far as the technology solution. When I was first introduced to it, I was excited by

Table 4.1 Differences between control towers and digital twins

Control tower	Digital twin
Consolidates near real time data	Additionally incorporates simulation data for predictive insights
Visibility primarily into logistics operations	Broader visibility including assets such as warehouses, transportation, manufacturing, suppliers and equipment
Track physical goods	Track physical goods and information flow
Focus on operational visibility	Focus, in addition to operational visibility, on planning, optimization and prediction including the cost and cash implications of the option analysis

the potential, however, it has not moved forward beyond visualization and is an application that overpromised and underdelivered. Although it represents a milestone in the evolution of visibility in the supply chain, it is a stepping stone to a future that is streamlined and simplified with an overarching user interface that users engage with to have visibility across the important elements within their ecosystem.

Blockchain

Blockchain technology traces its origins back to 1991 when Stuart Haber and W Scott Stornetta introduced a cryptographically secure chain of blocks to prevent document timestamp tampering. It came to prominence in 2009 when Satoshi Nakamoto (the pseudonymous developer[s]) devised the first blockchain database as part of the development of bitcoin.[20]

Blockchain provides a decentralized, secure method of recording transactions across multiple participants in a supply chain (e.g. supply, manufacturing and assembly processes) – see Figure 4.6. The benefits are transparency, less processing time among parties and a significant reduction in transaction costs. Using IoT, products are visible in near real time as they move through the supply chain and are securely recorded on the blockchain.

REAL-WORLD EXAMPLE William Grant & Sons, Ardnamurchan Distillery and Johnnie Walker

Whisky companies have been visibly embracing blockchain technology to build awareness and confidence with end customers by demonstrating transparency. William Grant & Sons launched the first ever whisky blockchain to ensure authenticity and traceability of its Ailsa Bay brand,[21] while the CEO of Ardnamurchan Distillery stated: 'the whisky production is a zero-sum process. What you get at the end must match the raw material's quantities. How can I proof precisely and transparently our whisky production? Blockchain.'[22] Furthermore, Johnnie Walker, with its blue brand, provides customers with the ability to scan a QR code to understand the journey the product takes from its source to them.[23]

Figure 4.6 Blockchain

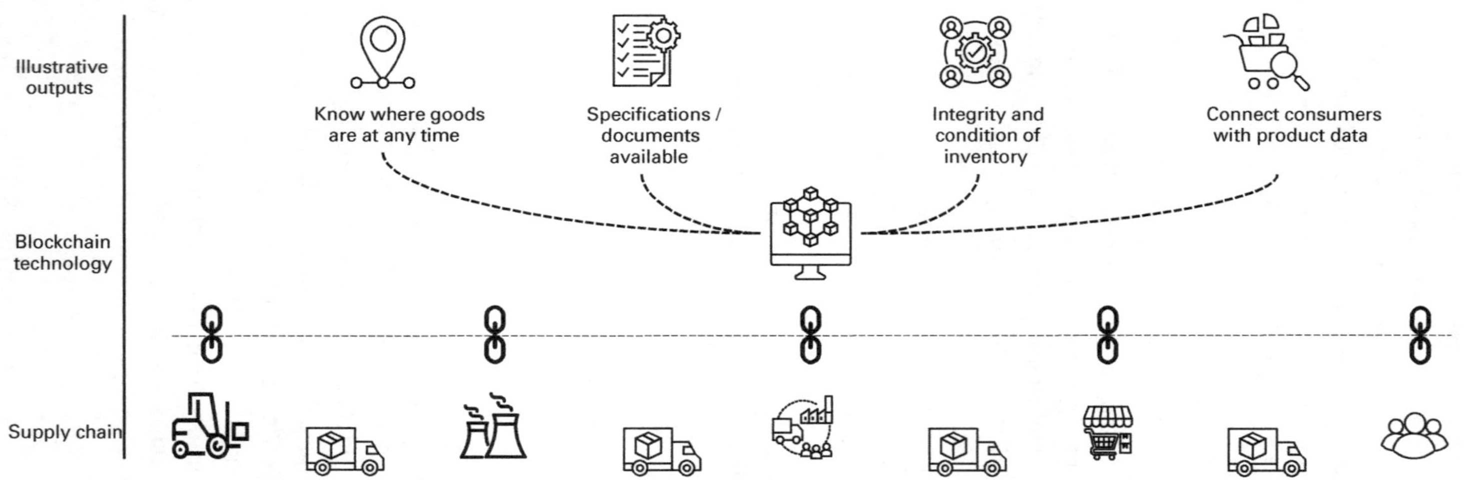

REAL-WORLD EXAMPLE IBM Food Trust

In 2017, IBM Food Trust was launched with 10 founding members, including Walmart and Nestle.[24] The blockchain solution was designed to enhance transparency and trust within the food supply chain. It addresses the increasing complexity and data growth in the food industry by providing a secure, permission-based platform where authorized users can access actionable food supply chain data. This includes the complete history and current location of any food item, along with accompanying information such as certifications, test data and temperature data. The blockchain technology ensures that all data is stored securely and is tamper-resistant, with access controls allowing data owners to maintain full control over who can access their information.

Within the solution is a governance model that includes policies to promote a collaborative ecosystem, with an advisory council overseeing the rules of engagement and accountability. The solution is built on the open-standard Hyperledger Fabric, which allows for interoperability with existing systems and scalability for future connections. The solution achieved success in providing trust in safer food, longer product shelf lives, reduced waste, faster traceability and better access to shared information.

Challenges with the use of blockchain start with how to integrate with existing systems because it needs to be able to record in a ledger what is going on with product(s) but also take updates from and provide updates to other systems such as ERP. Therefore, there is a potential risk of duplication of data and the need for vast computational power. Scalability is a challenge since the more users and transactions you have on a system increases the need to handle volume, which can lead to additional complexity. Also, there are privacy concerns in relation to sharing sensitive transaction data. These challenges provide reasoning for why this technology has not been embraced by many companies to date in their supply chain.

Digital product passport

A digital product passport (DPP) is a standardized, machine-readable record that contains detailed information about a specific product throughout its lifecycle. By 2027, DPPs will be mandatory for certain product categories, aiming to improve sustainability and transparency across the European Union.[25] A DPP serves as a unique identifier for a product, facilitating traceability, provenance, authenticity and interoperability (i.e. ability to exchange and make use of information across different systems) so that different supply chain stakeholders and consumers can be informed about the product. It creates a digital twin of a physical product, securely recording event, transactional and sustainability-based data, such as:

- material sourcing and manufacturing processes
- environmental impact
- composition and production history.

This digital twin is connected to the physical product via a QR code, barcode or other tagging solution, illustrating the product's sustainable, environmental and recyclability attributes. Figure 4.7 shows a digital product passport.

DPPs offer a wide range of benefits that can transform the way businesses operate and interact with consumers. By providing a vast array of reliable sustainability information, DPPs enable businesses to make data-driven decisions and reduce their environmental impact, while also allowing consumers to access detailed information about a product's origin, impact, materials used and reuse or recycling options through a simple scan. This increased transparency improves trust and loyalty among consumers, who can make informed purchasing decisions that align with their values. Furthermore, DPPs help businesses optimize their operations, reduce waste and streamline supply chains, leading to improved efficiencies and a more sustainable future.

As companies aim to meet the regulatory requirements of DPPs, they face several challenges. One of the first challenges is aggregating relevant information in a standardized format that all stakeholders can agree on and act upon. This can require additional investments, which could lead to higher prices for consumers. To overcome this, companies must prioritize data standardization and interoperability, ensuring that all parties can easily exchange and utilize DPP data. Another challenge is setting up an effective governance and project team for successful DPP implementation because DPPs impact many departments, such as sourcing, manufacturing, e-commerce and logistics. Therefore, it is crucial that the team has clear responsibilities and goals and is provided with training, so that the return on investment and visibility is enhanced.

DPP systems need to link with existing tools and systems, so the need for effective integration is key. Integration can be a challenge at the outset to

Figure 4.7 Digital product passport

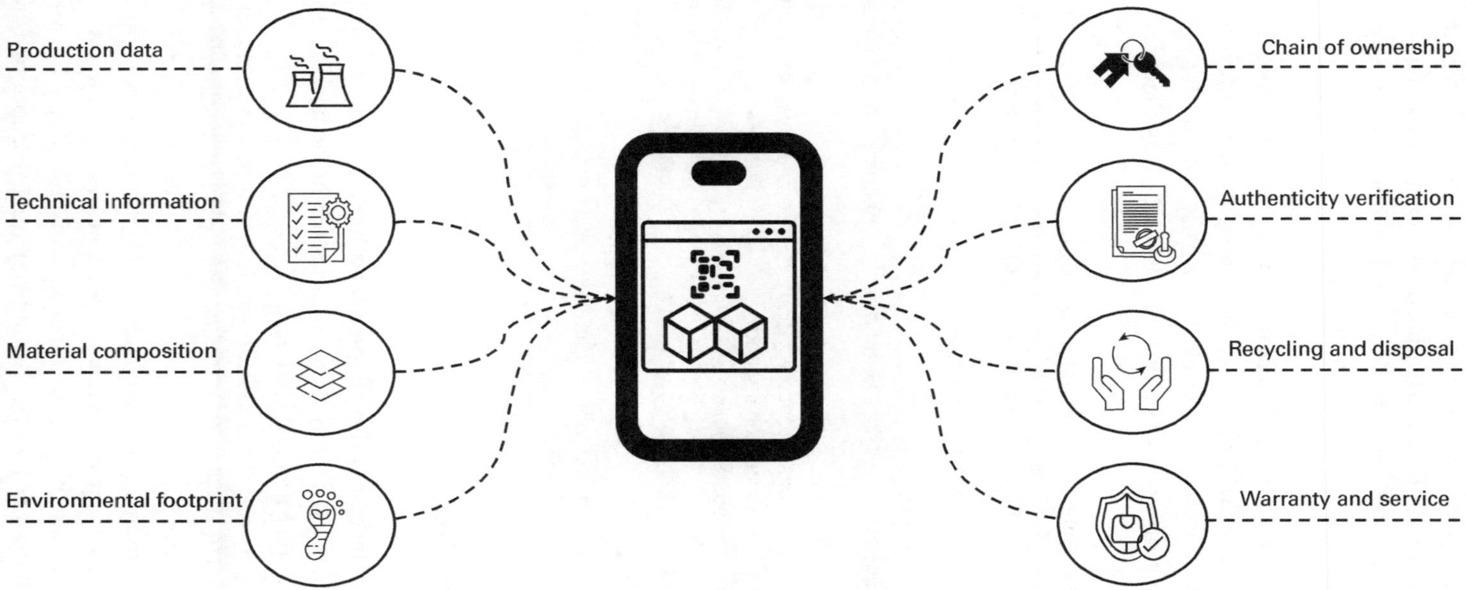

understand where data feeds will come from and whether they need a translation layer or can be directly pulled out of current systems and used.

Finally, the DPP solution should be continuously updated and maintained, so that companies can maintain the value of their products in their supply chain.

By addressing these challenges and implementing DPPs effectively, companies can achieve enhanced transparency, improved supply chain management and increased customer trust.[26] Table 4.2 outlines the sometimes-subtle differences between DPPs and blockchain.

Table 4.2 Differences between digital product passports and blockchain

	Digital product passport	Blockchain
Centralization	Centralized database	Decentralized, distributed ledger
Data storage	Stores data about a product	Records transactions and data
Purpose	Transparency, accountability and sustainability	Security, validation and immutability
Accessibility	Easily accessible and updatable	Accessible through a network of computers
Security	Dependent on database security	Uses cryptography for security

Conclusion

As we near the end of this chapter, there are a few key points that need reinforcing when considering opportunities for implementing digital technologies for visibility and transparency.

Firstly, near real time visibility enables companies to manage their inventory more effectively. Companies can reduce stockouts and overstocks through accurately forecasting demand and tracking inventory levels, leading to minimizing costs associated with excess stock and missed sales opportunities. Near real time monitoring can also significantly enhance risk management capabilities. Companies can proactively identify risks earlier through analysing data trends and potential disruptions before they escalate, allowing for timely interventions.

Digital technologies are improving communication and live data sharing among all stakeholders in the supply chain, leading to proactive problem-solving to address issues collaboratively, such as adjusting production schedules or sourcing alternatives. This also encourages alignment of objectives and work towards common goals leading to enhancing efficiency in the overall supply chain.

Having visibility and transparency assists companies to meet their sustainability targets from emissions to energy usage and environmental, social and governance (ESG) perspectives.

Finally, customer engagement and trust are improved by visibility and transparency. By providing near real time updates to customers throughout the 'order to fulfil' steps, they can see what is happening to their shipments and know the methods of sourcing and production used by the companies serving them.

What we are seeing is the emergence and rebadging of technologies that can help companies address the needs for visibility and transparency within their supply chains.

By embracing digital technologies, companies can not only streamline their supply chains but also respond more effectively to market demands, build stronger customer relationships and drive long-term success. As the supply chain continues to evolve, those who invest in visibility and transparency technology capabilities will have improved agility to handle the evolving needs of their customers.

Endnotes

1 Barrow, R (1979) 11th November edition of *San Diego Union* newspaper, https://sandiegouniontribune.newsbank.com (archived at https://perma.cc/624H-MRRE)

2 Hussein, M and Haddad, M (2021) US military presence around the world, Aljazeera, www.aljazeera.com/news/2021/9/10/infographic-us-military-presence-around-the-world-interactive (archived at https://perma.cc/2RYC-CANC)

3 Parlier, G (2011) Transforming US Army supply chains: Strategies for management innovation (Supply and Operations Management Collection) (Function), Chapter 18, *Business Expert Express*, 2011, Kindle edition

4 Parlier, G, Graham, D, Datta, S and Horowitz S.; Emails 8th and 31st August 2005 from Parlier, G (2011) Transforming US Army supply chains: Strategies for management innovation (Supply and Operations Management Collection) (Function), *Business Expert Express*, 2011, Kindle edition

5 Langford, C (2025) UK aiming to speed up delivery of new Challenger 3 tanks, *UK Defence Journal*, https://ukdefencejournal.org.uk/uk-aiming-to-speed-up-delivery-of-new-challenger-3-tanks/#:~:text="We%20are%20engaged%20with%20the%20supply%20chain,to%20meet%20the%20delivery%20timescales%2C"%20she%20said.&text=Trials%20are%20scheduled%20to%20continue%20through%202025%2C,and%20full%20operational%20capability%20(FOC)%20by%202030 (archived at https://perma.cc/55JU-LAAH)

6 Hammond, M F (2008) Sense and respond: Military logistics in a global security environment, *Army Logistician*, September–October 2008, https://alu.army.mil/alog/2008/sepoct08/pdf/alog_septoct08.pdf (archived at https://perma.cc/YQN2-VTJ5)

7 Rice, R (2020) From radar to RFID: How did we get here? *Control Design* www.controldesign.com/sensing/rfid/article/11298327/from-radar-to-rfid-how-did-we-get-here (archived at https://perma.cc/4DQ7-RU5X)

8 Nuffield Technologies (2024) 5 technologies that are transforming logistics in the military, https://nuffieldtechnologies.com/resources/blog/2024/01/16/logistics-military-five-transformative-technologies (archived at https://perma.cc/6RMG-TNJ5)

9 One Network Enterprises (n.d.) *Innovating the Defense Supply Chain*, One Network Enterprises, https://20608267.fs1.hubspotusercontent-na1.net/hubfs/20608267/industries/aerospace-defense/one_solution_defense.pdf (archived at https://perma.cc/2PXU-ZEF9)

10 Alicke, K and Foster, T (2024), Supply chains: Still vulnerable, McKinsey & Company, www.mckinsey.com/capabilities/operations/our-insights/supply-chain-risk-survey# (archived at https://perma.cc/WJU2-E9UU)

11 Gross, J (2025) Remember when a ship stuck for days came to be a metaphor for us all? *The New York Times*, March 16, Section F, p 6

12 Barnett, J and Fish, I (2024) How supply chain shocks have changed the face of global trade, *The Times*, August 27, Business and Money Section

13 Rizza, M N (2024) Worldwide enterprise resource planning applications market shares, 2023: Preparing for the AI-enabled world, IDC, https://my.idc.com/getdoc.jsp?containerId=US52728224 (archived at https://perma.cc/38CJ-494W)

14 Parametrix (2024) Cloud adoption among the Fortune 500, www.parametrixinsurance.com/blog/cloud-adoption-among-the-fortune-500 (archived at https://perma.cc/3MDJ-DAAW)

15 Radhakrishnan, S. (2021) How Walmart leverages IoT to keep your ice cream frozen, Walmart, https://corporate.walmart.com/news/2021/01/14/how-walmart-leverages-iot-to-keep-your-ice-cream-frozen (archived at https://perma.cc/HKP7-SMXF)

16 Caterpillar (2022) Inside the mind of a caterpillar data scientist, Caterpillar, www.caterpillar.com/en/news/caterpillarNews/2022/inside-the-mind-of-a-caterpillar-data-scientist.html (archived at https://perma.cc/GMS9-JMZZ)

17 Young, M, Soza-Parra, J and Circella, G (2022) The increase in online shopping during COVID-19: Who is responsible, will it last, and what does it mean for cities? *Regional Science Policy & Practice*, 14 (1), pp 162–179, ISSN 1757-7802

18 Siemens (2025) Digital twin, Siemens, www.sw.siemens.com/en-US/technology/digital-twin (archived at https://perma.cc/3RDB-LEET)

19 IBM (2024) Sweet success at British Sugar by embracing advanced technology, IBM, www.ibm.com/case-studies/british-sugar (archived at https://perma.cc/MUR9-HQCU)

20 IT Faculty (2018) History of blockchain, *ICAEW*, www.icaew.com/technical/technology/blockchain-and-cryptoassets/blockchain-articles/what-is-blockchain/history (archived at https://perma.cc/DVD3-BWEU)

21 Brett, C (2019) Ailsa Bay and its first blockchain whisky, *Enterprise Times*, www.enterprisetimes.co.uk/2019/12/02/ailsa-bay-and-its-first-blockchain-whisky (archived at https://perma.cc/G88A-XHV2)

22 Cedalio (2023) Supply chain: Revolutionizing a traditional industry with blockchain at Ardnamurchan, Cedalio, https://medium.com/@cedalio/supply-chain-revolutionizing-a-traditional-industry-with-blockchain-at-ardnamurchan-d795a9ea9d24 (archived at https://perma.cc/7AFU-725D)

23 Borsi, A (2023) Connected smart bottles by Diageo for Johnnie Walker Blue Whiskey, https://knowledge-hub.circle-economy.com/article/23341?n=Connected-Smart-Bottles-by-Diageo-for-Johnnie-Walker-Blue-Whiskey (archived at https://perma.cc/F262-CC3S)

24 Schilhabel, S, Sankaranarayanan, B, Basu, C, Madan, M, Glennan, C and McSherry, L (2023) Blockchain technology in the food supply chain: influences on supplier relationships and outcomes, *Issues in Information Systems*, 24 (3), pp321–332, https://iacis.org/iis/2023/3_iis_2023_321-332.pdf (archived at https://perma.cc/5QFS-QYCA)

25 Legardeur, J and Ospital, P (2024) Digital product passport for the textile sector, European Parliamentary Research Service, Scientific Foresight Unit, www.europarl.europa.eu/RegData/etudes/STUD/2024/757808/EPRS_STU(2024)757808_EN.pdf (archived at https://perma.cc/9TJB-X8VT)

26 Russo, A, Lopez, F, Cessel, A L, Carrere, R and Taupin, S (2024) Embracing Digital Product Passport as a regulatory requirement, Deloitte, www2.deloitte.com/us/en/pages/consumer-business/articles/embracing-digital-product-passport-regulatory-requirement.html (archived at https://perma.cc/Y65M-QK3J)

Leveraging data analytics for supply chain insights 05

In this chapter, we will look at the role of data analytics and how embracing it provides supply chains with more informed insights that improve decision-making. We will explore the use of predictive analytics and machine learning primarily and how these are being deployed by leading companies.

> **CHAPTER AIMS**
>
> - Uncover the significance of data analytics in driving supply chain insights and decision-making.
> - Learn how predictive analytics and machine learning enhance supply chain management.
> - Develop strategies for implementing data analytics to drive business outcomes.

'Information is the oil of the 21st century, and analytics is the combustion engine.'

PETER SONDERGAARD[1]

We have been hearing about big data and its use throughout the supply chain for nearly 20 years, but it has been a challenge for companies to prioritize it. Data is the bedrock of a company, yet it is an underappreciated asset and its use in the supply chain has been scattered. Recent studies have shown that 25 per cent of organizations have more than 50 distinct data silos; this shifts to 39 per cent if they classify themselves as highly data-driven.[2] This shows how challenging it is for companies to get to the data itself: having multiple systems equals multiple sources of truth, which leads to undermining and questioning what is the right answer when users are having to make decisions. Looking specifically at supply chain data, pin-pointing how much data is used to inform decisions is difficult to quantify; research is more attuned to *what* to use data for than to gauge *how much* is in use. When you step back and look at how complex global supply chains have become, with increasingly interconnected multifaceted needs, the ability to analyse vast amounts of data from multiple data sources beyond the traditional inputs of historical data (via ERPs, IoT, unstructured variable data) should form the foundation of decision-making. It will lead to operational efficiency and in time, provide competitive advantage. As companies endeavour to be at the forefront of customers' minds and pockets, the need for greater agility, efficiency and responsiveness should not be ignored. Data analytics integrated with emerging technologies, such as predictive analytics and machine learning, provide indispensable tools for amplifying supply chain performance.

EXPERT INSIGHTS

The following insights have been taken from speaking with an expert who was heavily involved with Unilever through significant change in relation to supply chain analytics.

Unilever is one of the world's largest consumer goods companies, established over 100 years ago. In 2024, its turnover was €60 billion, it was present in 190 countries, with 280 factories and 57,000 suppliers,[3] 160,000 unique materials from 8,000 suppliers, delivering to over 450 distribution centres and supplying approximately 250,000 unique customer locations. It was recognized as 'Masters' in supply chain by its peers for a seventh consecutive year as part of Gartner's annual supply chain review in June 2025.[4]

Back in 2017, it decided that its key priority was to utilize data and analytics in the supply chain, leading to a significant programme of investment over the following years. The journey began with a strategic decision to understand the role of the supply chain and to embrace a digital mindset. The initiative was driven by the recognition that many of its talented staff had been working in the supply chain for a considerable time and may not be equipped with the relevant skills to handle the complexities of a modern, digitized one. Unilever aimed to upskill this talent pool as well as bring in new talent to act as a catalyst for its supply chain transformation, with a vision of how it wanted to work going forward.

Its transformation was guided by three emerging themes: hyper-fragmentation, customization and personalization. **Hyper-fragmentation** referred to the increasing complexity it had in its product offerings, which was only becoming more complex. For instance, a single product category such as mayonnaise, which traditionally had only a couple of stock-keeping units (SKUs), was expanding into hundreds of variants due to packing types, flavours and formulations. Trying to manage this expansion as it had done in the past was very difficult without being able to get data into the hands of users in a timely manner to support decision-making. Forecasting complexity had dramatically increased. Aggregation was not now an option for forecasting with accuracy, with there being so many nuances and it required a more sophisticated supply chain system capable of managing numerous SKUs efficiently. In addition to that was the usual complexity that comes from product life cycle management and new product introduction.

Customization of products was driven by the demands of Unilever's customers, who were primarily retailers and distributors. However, it was also touching new consumers with the emergence of direct-to-consumer channels in some markets. These different stakeholders sought unique products tailored to their specific needs. To differentiate between them and their competitors, Unilever needed a different approach to manufacturing. In the past, long production lines had been the norm but now smaller productions that were customized with more wash-downs and sequencing required supply chain analytics to help determine optimal approaches.

Finally, **personalization** (while aspirational in 2017 but would become more important in time) aimed to personalize products. This was more focused towards health and beauty sectors, such as personalized skincare. Unilever's challenge was how to gauge the level of personalization from customers two steps removed from it in the supply chain. The customer would engage with a retailer; however, the signal for replenishing stock could come from an IoT device on the retailer's shelf and then trigger a demand signal to Unilever. The intricacies of data sharing and the need to be agile and responsive led it to challenge how it was set up in the supply chain. It also had to be responsive with retailers who were selling it space on shelves, which needed to be replenished frequently.

To focus on these three themes, a digital leadership team for the supply chain was created. The team was composed of senior leaders tasked with driving the transformation. Various workstreams were initiated to separate the work across the supply chain:

- Assessing how technology was used to improve forecasting processes.
- Reviewing the supply processes and how they engaged with the supply base.
- Assessing what manufacturing capability was required versus contract manufacturing (i.e. contract and instruct another company to manufacture on behalf of Unilever).
- Rebuilding manufacturing capabilities.
- Digitizing warehouse operations, including the introduction of automation and robotics (which we will discuss in more detail in the next chapter).
- Serving customers.

The transformation was approached by a two-stage process of consolidating data into a data lake to provide near real time visibility of the supply chain and then deploying decision intelligence via advanced analytics and AI.

Global process owners, the custodians that own the end-to-end processes across the supply chain function for all business units globally, were put in place. At the heart of these changes was the urgent need for data and analytics to underpin the value chain, with the creation of a data lake to centralize and standardize data.

The governance of the transformation involved multiple sponsor groups, including the digital supply chain team, the business services team and the integrated operations (IOPS) programme. Each group had its own objectives but relied on the data and analytics infrastructure to achieve its goals. This alignment of interests helped drive the transformation forward, despite the inherent challenges of transforming the global supply chain with its existing data architecture.

Moving a company from a local perspective to a global one led to initiatives appearing locally with IT departments that had built bespoke systems to meet their specific needs. This, in time, further complicated the integration process and acted as 'plasters', since the need to solve local issues took priority within a country or region, instead of focusing on the overall goal. Data was often transformed for specific purposes, making it a challenge to achieve a single version of the truth. This was compounded with timing issues, which naturally arose due to the global nature of the business: data was processed at different times in different regions across four ERP systems, leading to inconsistencies and user confidence issues at various stages of the transformation. The implementation of the data lake, connecting over 40 systems with over 750 data pipelines, aimed to address these and other challenges by providing a centralized consistent source of data. However, the process was hindered by a lack of detailed design knowledge among global process owners and the difficulty of integrating data from many different systems and instances of the same systems.

Despite these challenges, there were notable successes along the journey due to having to adapt to upgrades and refreshing technology capability. Fortuitously, the data lake and visibility solutions became available as Covid-19 happened, so were immediately put into use and became central to daily tactical decision-making that supply chain leaders had to make to successfully navigate this period. Without that investment, there wouldn't have been visibility, which would have hampered the ability to respond as well as Unilever did. Another success was in the procurement transformation, which resulted in the creation of a procurement 'console' that provided comprehensive data access for sourcing and supply management decisions. This initiative alone was recognized externally with awards[5] and demonstrated material benefits.[6]

Similarly, forecasting improvements were achieved through enhanced analytics architecture by digesting structured forecast data and analysing it in a consistent manner. This has moved the dial from forecasting successes in the mid-50th percentile to 80 percentiles.[7]

The overall impact on supply chain performance was mixed. While the digital transformation enabled stronger data visibility and decision-making capabilities, the full potential of these changes was not always realized due to resistance from people in roles most affected by the new systems and capabilities. The accuracy and scalability of the analytics were often questioned, leading to delays in adoption or users still manually overriding them. This is a challenge many companies face moving towards autonomous planning, since the user will want to make decisions for other reasons beyond what the data solely is telling them – for example, based on intelligence gathered from commercial teams and marketing campaigns (the theme of trust of technology will be revisited several times again in the book).

In conclusion, Unilever's journey in supply chain analytics over the period discussed here was marked by a significant strategic shift, with the introduction of new talent and the implementation of advanced data and analytics systems. While there were notable successes and they continue this journey, the complexity of global supply chains, resistance to change and the need to continue to evolve systems brings to bear the importance of data and analytics and its role in helping companies prosper. Unilever highlights the importance of detailed design knowledge, centralized data management with clear ownership within the supply chain and the need for continuous adaptation to realize the full benefits of digital transformation in supply chain management.

The purpose of data analytics

The purpose of data analytics to supply chain is to metamorphose raw data into meaningful insights that guides users to informed decision-making. Supply chains are multilayered, frequently spanning the world through nth tier suppliers and a variety of transportation modes. This complexity requires innovative approaches to processing and analysing data to identify trends, mitigate known and unknown risks and optimize a multitude of elements of the supply chain.

Analytics for far too long has been shining insights from a descriptive perspective, staring at what has happened in a cobbled-together slide deck or static visualization tool, based on whatever information is deemed important at the time. Yet, running supply chains looking backwards all the time is unhelpful. Analytics can be a game-changer when it is deployed to provide prescriptive and predictive capabilities that aid the user to improve decision-making and to accelerate planning. When data is regarded with respect, maintained and then integrated with analytics, it empowers companies to be proactive, which will materially improve service, cost and capital positions.[8]

It isn't a stretch to suggest data analytics is the backbone of an intelligent supply chain – one that can access, process and analyse data to make informed decisions at each stage.[9] From forecasting, procurement, production and inventory management, through to distribution, data analytics provides users with insights that support continuous improvement. It also helps with visibility, improves collaboration from end-to-end and enables risk mitigation strategies. When used appropriately, it will enable acceleration of decision-making by combining data quickly. Supply chain analytics has been deemed for a while now to be one of the most important emerging technologies, as per supply chain practitioners,[10] who have voted it the fourth most important trend out of the last six years. However, despite it sounding like a 'silver bullet', it remains untapped in many instances.

Data analytics has various stages of maturity as shown in Figure 5.1.

Figure 5.1 The different stages of data analytics

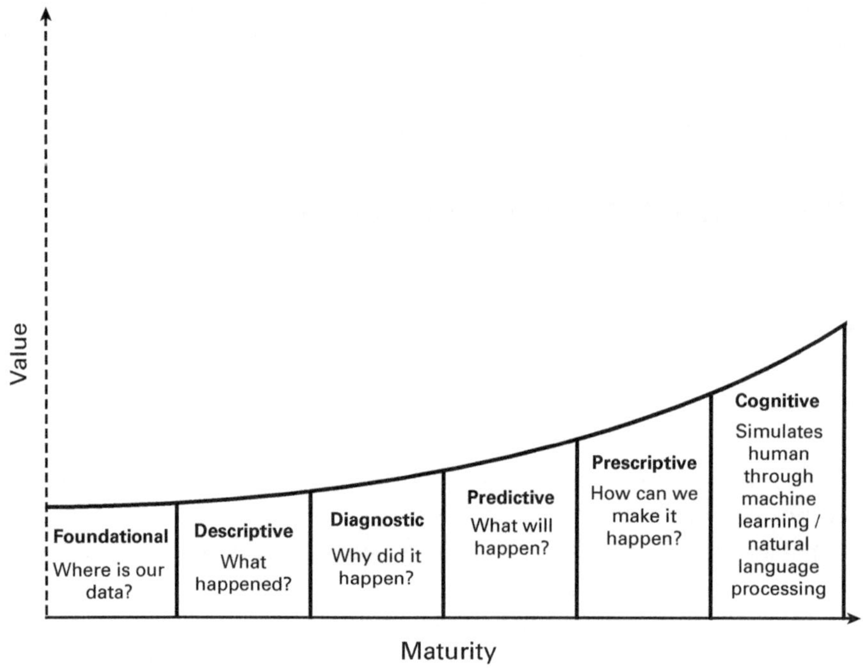

Data analytics starts with **descriptive analytics:** this is what has been the most common and easiest to collate. For this, you are backward-looking to see what happened to inform future direction. This is achieved via analysing historical data to understand past trends and patterns and then converting these into data visualization tools via reporting and dashboards, as examples. However, many companies are reliant on dashboards that never refresh what they look at; to quote Rollo May, 'dashboards are not windscreens, they are rear-view mirrors.'[11] If your company is at this stage of maturity, my suggestion would be to agree on what the key reports and dashboards you need to run the supply chain are (led by CSCO and their leadership team) and then formalize these to move forward to the second stage.

The second stage is **diagnostic analytics**, which is focused on showing why something happened and getting to the root cause of a trend or pattern. This can be done by using drill-down analysis, correlation analysis and data mining, as examples.

Moving on from using analytics for static viewing, the third stage is **predictive analytics**, which is where you move into future casting and try to determine what will happen. This is when the use of statistical models and machine learning algorithms are deployed to provide confidence and support decision-making using, for example, clustering, decision trees and regression analysis.

> Let's look at clustering briefly. It is a data mining technique that groups similar data points together, and cluster analysis automatically categorizes products based on factors (e.g. fast moving, margin, shop coverage), which then allows you to see distinct groups. I worked with a UK retailer who had over 18,000 products that it sold primarily via its own store estate. The ability to get a handle on inventory positions was challenging, which led in many instances to stock outs or holding considerable inventory (sometimes leading to storing 'seasonal' stock until the following year, such as Halloween or Christmas). Using the clustering approach, we were able to get a handle on inventory – not by category but by grouping products that had similar characteristics. Figure 5.2 shows the clustering process.

Figure 5.2 Clustering process

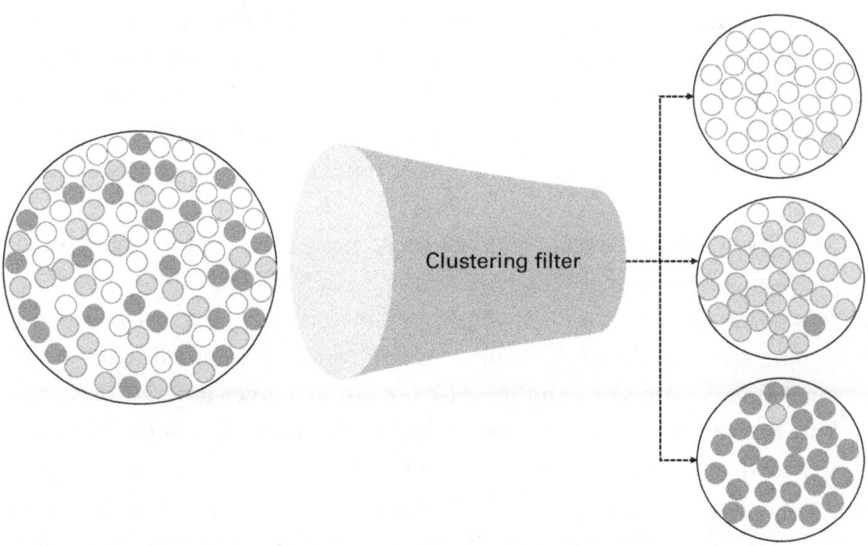

The clustering data was refreshed regularly to aid the inventory team via static visual dashboards. This meant the retailer could improve its decision-making on what products to focus on, leading to creating an opportunity to reduce working capital by approximately a quarter.

The fourth stage is **prescriptive**. This is the stage where analytics assists users to answer the question of what they should do using advanced methods such as optimization techniques. Examples of this would include using decision analysis, linear programming and simulation.

Explainable analytics is the fifth stage. This is where the user can understand why analytics is stating what it is and why the recommendation should be trusted. Examples include model interpretability and partial dependence plots.

The final stage is **autonomous analytics** where decisions are made using AI and machine learning to automate analytics and decision-making process. Examples include automated reporting, near real time analytics and self-service analytics.

Data analytics in action in the supply chain

There have been significant advancements made in predictive analytics and machine learning for the handling of data.

As I touched on before, predictive analytics, as the name suggests, uses historical data to predict future trends and events and inform strategic direction.[12] In supply chains, this form of analytics can assist users to anticipate changes in demand, identify potential disruptions and proactively adjust strategy. The main difference is it moves supply chain insights from gazing at past performance to anticipating future trends, which supports informed decision-making.

Machine learning (ML) as outlined by Arthur Samuel, a forefather of AI, is a subset of AI that enables computers to learn from data sets without explicitly being programmed.[13] In supply chains, ML algorithms can analyse large data sets and identify complex patterns, offering companies the ability to quickly make decisions and increase operational efficiency.

One of the primary benefits of deploying the above techniques for data analytics is visibility, which we explored in detail in the last chapter. Leveraging technologies like IoT, radio frequency identification (RFID) tags and global positioning system (GPS) tracking assists companies to monitor goods, assets and shipments through the end-to-end supply chain in near real time. This helps to identify potential disruptions and inefficiencies, enabling companies to enact corrective action before problems occur or escalate. For example, Kroger in October 2024[14] announced its use of RFID to track shipments and deliveries of bakery products with data analytics enabling it to proactively manage freshness and inventory levels.

An ongoing theme seen in many large retailers (e.g. Ocado, Tesco, Walmart) has been deploying data analytics to help them predict stock outs and then balance their supply chains to minimize these across their store portfolio. Data analytics predicts surges in demand at different levels such as country, region, county and city level, rebalancing the supply chain.[15]

Planning internally – from the demand forecast, through the supply chain, to production and beyond – all involve complex decision-making. Data analytics aids users in determining trade-offs between demand, supply and production. Predictive models can anticipate bottlenecks while algorithms can optimize sequencing and scheduling of production runs, for example. L'Oreal improved its forecast demand by exposing different forms of data (structured and unstructured) to analytics to propose the best decisions to demand planners in a timely manner.[16]

The ability to handle different forms of data has only been a reality since 2018. From analysing past sales patterns and then using mathematical techniques to improve the forecast, we have made several leaps forward to a time now where we can inform forecasts with many external factors such as weather, local events, economic indicators and social media, meaning a more accurate demand forecast to feed out throughout the supply chain to improve its overall performance.

However, we need to act with caution when suggesting that implementing analytics to traditional demand planning techniques is the answer alone. Jonathan Karelse,[17] in his book *Histories of the Future*, says many times throughout that forecasting is inherently wrong and to suggest otherwise is redundant thinking. To put it another way, if analytics alone could solve forecasting, then why aren't people using it to predict and win the lottery? Where analytics is beneficial is to save users' time: rather than spending considerable time cleaning up data and evaluating it using only one or two variables, the decisions generated by analytics accelerates the critical thinking process. Also, analytics helps supply chain planners move beyond treating all categories and SKUs the same. For example, if a planner for a variety of categories exposes each to the same variable datasets, mistakes will happen since each category requires datasets that directly correlate to it and affect it.

An area where data analytics is also playing a beneficial and important role is by aiding users with 'what if' analysis. This is where options are provided to assess SKUs using specific variables (e.g. What if we ran a marketing campaign? What if a competitor reduces its prices?) rather than all or no variables. If a demand planner is looking at hundreds of SKUs, supply chain leaders can focus their attention on the critical high-performing SKUs and take recommendations from the analytics about the tail SKUs (e.g. slow-moving, unpredictable demand patterns, less frequently replenished SKUs) and an example of this is shown in Figure 5.3.

Surprisingly, few organizations use Forecast Value Analysis (FVA) as part of their review process of forecasting, but it can add value in 'what if' analysis as well. FVA, in simple terms, shows you what effect (positive or negative) an intervention has on a forecast and over time, helps users and algorithms understanding so as to improve the forecast.[18] FVA was developed in the early 2000s by the forecasting guru Michael Gilliland.[19]

Balancing inventory management is an example of when data analytics comes into its own. The pain of over- or under-ordering products leads to the dichotomy of having excess inventory, which ties up working capital or stock outs and results in lost sales and disgruntled customers. Predictive analytics assists the user in making sense of inventory, helping them optimize it by forecasting demand better and identifying the best cost to serve approach to storing and moving products. For example, Nike[20] uses advanced algorithms to monitor inventory levels across its entire distribution

Figure 5.3 Inventory analysis graph

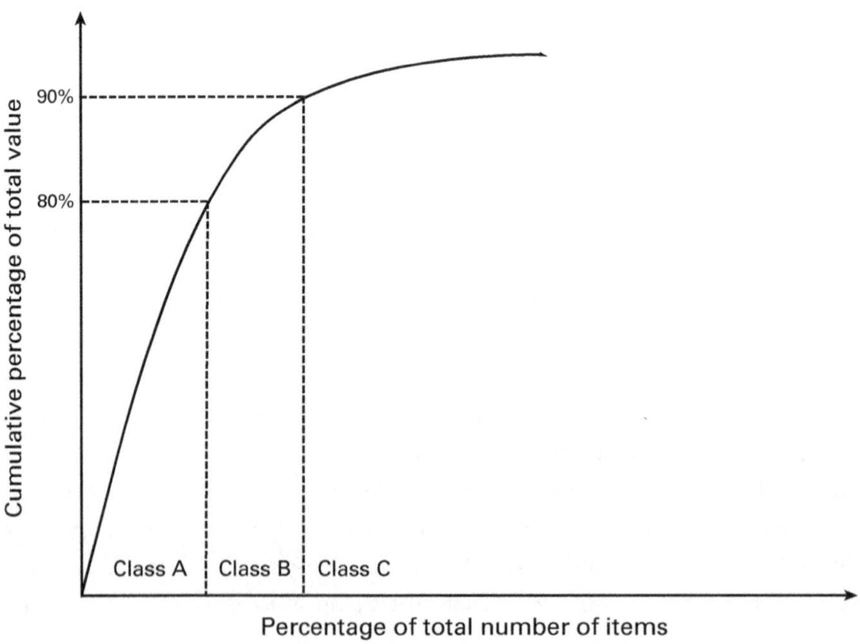

network. By analysing historical demand patterns, it can forecast future inventory needs, minimizing both overstocking and shortages, managing the supply through its channels (store, 'click and collect' and online) therefore creating opportunity to meet demand with flexibility.

Moving beyond capabilities that analyse your own part of the supply chain ecosystem, data analytics can help you monitor supplier performance and manage your company's supply chain risk profile – developing predictive analytics to track supplier performance in near real time, by assessing factors such as supplier financial stability, timeliness and quality of products they provide. P&G[21] uses data analytics to monitor its supply base's financial health. By examining supplier credit ratings and historical performance data, it can predict potential supplier failures and develop contingency plans to minimize disruption to continuity of supply.

Resilience is a supply chain trend that is usually spoken about after exogenous shocks (e.g. natural disasters, geopolitical events and market volatility), which have exposed a company's lack of it! Resilience has grown significantly in importance during and post the Covid pandemic. Prior to then, companies typically looked at resilience by conducting one-off exercises, mapping a finished good back through tier 1 (and possibly tier 2) supply base and understanding the potential risks. However, with the emergence of analytics and technology tools, this can now be managed proactively in near real time to help users determine how they want to evolve their supply chain. By continuously analysing data, companies can detect early signs of supply chain delays or supplier failures, leading to quicker decision-making and more effective mitigation strategies. For example, a US automotive parts supplier successfully built resilience through conducting a risk assessment of its supply chain to reduce dependence on China, which led to it expanding its supplier base and transferring up to 25 per cent of its sourcing to new, diversified locations.[22]

Also, companies, such as Resilinc,[23] provide solutions for clients to map their extended supply chains on its platform, amalgamating data from a company's supply chain (e.g. supplier plant locations and spend, company assets, customer locations, route to markets) with a variety of external sources (e.g. news feeds, weather, geopolitical factors). This is to determine risk profiles and help users see their risk profile at any given time. Working with their supply base, companies can quickly assess where they have exposure to risk and then proactively manage it to minimize deploying strategies such as dual sourcing or working with suppliers to develop alternative manufacturing locations. Toyota deploys such an approach; it views stability in its supply base as key, being transparent and sharing, it uses analytics to keep category resilience high.[24]

Predictive analytics originally entered the supply chain in relation to asset management and is still a popular use for it. Using analytics, companies can understand machine performance data, therefore determining when the machine is likely to fail or require servicing to minimize costly downtime. Reckitt, a UK multinational consumer goods company known for its continuous improvement mindset, has been embracing analytics to help it in its global operations. In one of its plants in the UK, it was observing 30 per cent down time due to machine issues. Through implementing a condition-based maintenance approach (which uses sensors and other monitoring equipment to collect data), it was able to analyse data to identify patterns and anomalies to help better inform how to reduce machine down time, 'virtually eliminating unexpected failures and extensive machine related time'.[25]

Transportation is typically a significant part of supply chain expenditure. Having data analytics capabilities helps users assess routing to optimize it based on various variables and manage their fleets more efficiently. By assessing near-real time data such as traffic, weather conditions, fuel costs and delivery requirements, companies can make informed decisions on how to move their products more effectively. For example, Amazon Logistics utilizes 20 machine learning models[26] with a variety of variables including distance, fuel prices, weather and expected delivery times to optimize delivery routes. This helps to reduce transportation costs and improve delivery performance.

REAL-WORLD EXAMPLE Formula 1

When you think of Formula 1, supply chain analytics is probably not the first thing that comes to mind. However, with 24 races across five continents and 21 countries in nine months, the operation is complex. There is predictability in knowing where the cars must be at any time, but when you break down the car into over 14,000 components[27] that are unique to each team, you then must know when you will need to replace items based on what is happening to them at each race. This is where the need for analytics comes to the forefront. From using predictive analytics for wear and tear of components and monitoring performance, reacting to what is going to happen (as well as what is happening) requires teams to be on top of their data to inform decision-making. With transportation of personnel and equipment, analytics is used to determine optimization of schedules, which transport mode to use and to determine the best route to pick up components when they are needed. Teams will use hubs to store inventory for different legs; for example, most teams during the European leg will have a central warehouse to service. Once that leg is complete, a new warehouse needs to be set up, so inventory moves with the team. It is estimated that annually, each team spends approximately $100m[28] doing this. Therefore, if they don't execute it correctly, the cost is significant. Analytics is also important from a sustainability perspective: if components are regularly in transit, consideration needs to be given to carbon emissions and how to operate at the lowest level of emissions. This has led to engagement with logistics providers and using electric vehicles within the supply chain with the goal of being 'net zero carbon' by 2030, set by the International Racing Body. From 2018 to 2022, the F1 supply chain had already decreased this by 13 per cent.[29]

This section has laid out a few examples of where data analytics can be deployed to support users with intelligence in the supply chain. What the Formula 1 example demonstrates is that there are multiple benefits to using data analytics to improve supply chain performance. While companies should be cautious about being over-reliant on data, believing it is the answer to all, it helps users quickly get to a decision that is more informed, compared to traditional methods.

The tricky part is how to implement an analytics approach that can quickly provide benefit and not require an overhaul of legacy technology. The next section details strategies to consider as you move your company forward with using the currency of data to empower your supply chain.

Strategies for implementing data analytics

To leverage data analytics to drive improved business outcomes, companies must adopt a structured approach and make this repeatable. This requires companies to integrate tools to develop a data-driven culture that continuously iterates and refines strategies to answer the questions imperative for improving their performance. Below are the key strategies that will help companies achieve successful implementation of data analytics in their supply chains.

Data collection and integration

The first and most important step is to bring data together from disparate sources and have it at a level of quality where it can be amalgamated to provide a comprehensive view of the supply chain. Disparate sources are a challenge as many companies have moved away from managing their supply chain data primarily through the enterprise resource planning (ERP) system and complementary supporting modules. When I started work in the early 2000s, it was typical for a company to have an ERP (in my case, it was the AS400 system) that did most of the transactions for supply chain activities, from holding demand data, sales orders, bill of materials (BoM), purchase orders etc. Nowadays (said with relief), a typical manufacturing company will have ERP but will also have an advanced planning system (APS), manufacturing execution system (MES), warehouse management system (WMS) and transportation management system (TMS), for example, among other systems which could all be provided by different vendors. Therefore, the logic, data and fields could be vastly different.

Therefore, the need for integration of data is critical to move the company beyond siloed time-consuming ways of working through the various stages of the supply chain. Deploying enterprise data platforms or cloud-based solutions can help centralize data for access and analysis. However, this is a time-consuming and expensive process – think of a data lake fed by all the systems – and although this is still common practice, a new technology (i.e. Generative AI or one of its future iterations) could transform this process, which I will discuss in Chapter 13.

As well as integration, the elephant in the room is the quality of data. It is easy to say that your data needs to be accurate, clean and consistent. However, in practice, changing user mindsets to see the importance of keeping data this way is a real challenge. This has led to the implementation of data governance policies to address data integrity issues, as well as companies upgrading their ERP systems, to use that as a justification to get data in better shape. However, for data to be high quality and an asset, it needs ongoing maintenance and tight controls over who can edit it and the data governance parameters.

In large multinational companies, the role of global process owners has become popular. Moving forward, I believe this role should change to 'global process and data owners' due to the increasing integration of data into important processes and common standard approaches to managing each of them. Many professionals in the supply chain community suggest data belongs to IT and vice versa. With processes leading the way, the data that underpins it is the crucial element.

Adopting advanced analytics tools

Investments in tools and platforms to support data analytics is critical to move your company forward. These will provide you with a range of functions, such as descriptive analytics, machine learning, predictive modelling and visualization abilities.

Solutions should be cloud-based since they will provide the company with the scalability and flexibility required for handling the volume of data required to support analytics. These cloud-based solutions should be coupled with tools that can use AI and machine learning to build models for tasks across the supply chain, from demand forecasting to predictive maintenance, to route optimization.

Collaboration

Building a mindset that openly embraces analytics cannot be achieved without collaboration from departments beyond the supply chain function, which can be challenging for many companies. Without engaging with other functions such as IT, commercial and finance, you risk only looking at problems from a narrow perspective.

The company culture needs to be nurtured towards one that considers itself data-driven. This requires time to define and embed where leadership needs to be educated on the value of data and analytics and then through messaging and actions provide investment and time to employees to upskill. The importance of leadership buy-in cannot be underestimated and its alignment with the broader business goals will create excitement for personnel to adopt this new approach to data.

Continuous improvement mindset

Data analytics only works when it is an ongoing process. It should not be static and if you have users 'bought in' to the culture, they should be challenging the business to drive improved insights regularly. By monitoring, testing and iterating with data and visualization views, the users will gain confidence in the data.

Assessment and refinement of models and algorithms should be continuously occurring as new data is made available. Experimentation should also be regularly happening, for example, running A/B tests to evaluate how effective new strategies and models are performing. In time, this should uncover greater insights and further opportunities.

Conclusion

Data analytics and its use in supply chain should not be underestimated. While the path to it is challenging, with the obvious hurdle of poor-quality data always being held up as the reason why it cannot work, companies need to move beyond this outdated view and embrace data and the ability to analyse it, to aid them in their mission. Data analytics in different forms are empowering companies to run their supply chains from a primarily data-driven approach, as shown with the company examples in this chapter.

Through leveraging analytics, companies can reduce costs and better mitigate risks, providing competitive advantage and increasing customer satisfaction. Investment in supply chain technology and capabilities is a must for companies that want to be leaders. Those who don't embrace a data-first approach will continue to miss out on opportunities to enhance their supply chain and overall business performance.

Endnotes

1 Sondergaard, P (2011) Gartner Symposium / ITxpo, October 2011, Gartner
2 Saran, C (2020) Data silos and IT complexity stifle business potential, *Computer Weekly*, www.computerweekly.com/news/252492360/Data-silos-and-IT-complexity-stifle-business-potential (archived at https://perma.cc/2TDC-S6W9)
3 Unilever (2025) Unilever at a glance, www.unilever.com/our-company/at-a-glance (archived at https://perma.cc/W8JZ-WVH4)
4 Gartner (2025) Gartner announces 2025 rankings of the Global Supply Chain Top 25, 18th June 2025, www.gartner.com/en/newsroom/2025-06-18-gartner-announces-2025-rankings-of-the-global-supply-chain-top-25 (archived at https://perma.cc/F56L-XGRF)
5 CIPS (2025) Procurement Excellence Award, https://excellenceawardscips.com/2025/en/page/home (archived at https://perma.cc/VBP7-HJGF)
6 Cosgrove, E (2020) Unilever CSCO: Agility beats forecasting when the supply chain is stressed, Supply Chain Dive, www.supplychaindive.com/news/unilever-csco-agility-forecasting-coronavirus/581323 (archived at https://perma.cc/2GGX-P5UV)
7 Karwal, A (2023) Transcending boundaries: Unilever's cash flow forecasting foresight, *EuroFinance*, www.eurofinance.com/news/transcending-boundaries-unilevers-cash-flow-forecasting-foresight/#:~:text=The%20implementation%20has%20considerably%20improved,at%20the%20three%2Dmonth%20horizon (archived at https://perma.cc/8BEB-75FU)

8 Abdullah, O (2021) Supply chains must be proactive and reactive. Here's how to do both, Supply Chain Brain, www.supplychainbrain.com/blogs/1-think-tank/post/33025-supply-chains-must-be-proactive-and-reactive-heres-how-to-do-both (archived at https://perma.cc/R5WC-SEXL)

9 Raj, A, Kumar, R R, Narayanan, S, Kirca, A H and Jeyaraj, A (2025) Analytics capability, supply chain capabilities, and operational performance: A meta-analytic investigation, *Journal of Operations Management*, https://onlinelibrary.wiley.com/doi/10.1002/joom.1376 (archived at https://perma.cc/ZN63-XST4)

10 ASCM (2025) Top 10 supply chain trends, ASCM, www.ascm.org/making-an-impact/research/top-10-supply-chain-trends-in-2025 (archived at https://perma.cc/2A4H-6YX5)

11 May, R (1994) *The Courage to Create Reissue*, W W Norton & Company

12 Cote, C (2021) What is predictive analytics? Harvard Business School, 26th October, https://online.hbs.edu/blog/post/predictive-analytics (archived at https://perma.cc/Z3SF-47DG)

13 Brown, S (2021) Machine learning, explained, MIT Management Sloan School, 21st April, https://mitsloan.mit.edu/ideas-made-to-matter/machine-learning-explained (archived at https://perma.cc/B9V9-VF86)

14 Meyer, G (2024) Kroger adopts RFID to maximize grocery freshness and reduce waste, IMPINJ Insights, 11th November, www.impinj.com/library/blog/kroger-adopts-rfid-to-maximize-grocery-freshness-and-reduce-waste (archived at https://perma.cc/XD2D-AQZV)

15 Retail Week (2023) Amazon and Tesco have 'best in class' supply chain strategies, *Retail Week*, 19th April, www.retail-week.com/supply-chain/amazon-and-tesco-have-best-in-class-supply-chain-strategies/7043552.article (archived at https://perma.cc/PXB6-PCJL)

16 Seifert, R W and Markoff, R (2022) L'Oreal: the beauty of supply chain digitalization, I by IMD, *IMD*, 13th July, www.imd.org/ibyimd/innovation/loreal-the-beauty-of-supply-chain-digitalization (archived at https://perma.cc/H42L-UB54)

17 Karelse, J P (2022) *Histories of the Future*, Forbes Books

18 Chase, C (2021) Assisted demand planning using machine learning, pp 111–114, *Business Forecasting*, Wiley

19 Gilliland, M (2002) Is forecasting a waste of time?, *Supply Chain Management Review*, July–August, https://www.thefreelibrary.com/Is+forecasting+a+waste+of+time%3f+(Forecasting).-a089076989 (archived at https://perma.cc/XF5M-PPUF)

20 Henrich, J, Li, J, Mazuera, C and Perez, F (2022) Future-proofing the supply chain, McKinsey, www.mckinsey.com/capabilities/operations/our-insights/future-proofing-the-supply-chain (archived at https://perma.cc/8XNK-DGG9)

21 Knime (2022) How P&G uses real-time data for supply chain resiliency, Knime, www.knime.com/success-story/how-pg-uses-real-time-data-supply-chain-resiliency (archived at https://perma.cc/Z5R2-LJ9E)

22 Catherine (2022) Real-world supply chain resilience, Transformeus, https://transformeus.com/real-world-supply-chain-resilience (archived at https://perma.cc/4E87-Y6NM)

23 Mizell-Pleasant, A (2025) Supply chain industry sees nearly 40% annual increase in disruptions: Research, Supply & Demand Chain Executive, www.sdcexec.com/software-technology/supply-chain-visibility/news/22931190/resilinc-supply-chain-industry-sees-nearly-40-annual-increase-in-disruptions-research (archived at https://perma.cc/LBG5-9UKE)

24 Shih, W C (2022) What really makes Toyota's production system resilient, *Harvard Business Review*, https://hbr.org/2022/11/what-really-makes-toyotas-production-system-resilient (archived at https://perma.cc/C33L-PJXJ)

25 Ellins, B (2023) Factory of the Future and Digital Transformation, 7th June 2023 Presentation, Gartner Supply Chain Symposium / Xpo, Europe

26 Tarasov, K (2024) Amazon is using generative AI to drive more same-day shipping using smarter robots and better routes, CNBC, www.cnbc.com/2024/09/17/how-amazon-is-using-generative-ai-to-drive-more-same-day-deliveries.html (archived at https://perma.cc/JD45-AT9A)

27 Motorsport (2023) Insider's guide: How is an F1 car made? Motorsport, www.motorsport.com/f1/news/how-is-an-f1-car-made/7626323/#:~:text=How%20many%20parts%20is%20an,hand%20processing%20to%20build%20them (archived at https://perma.cc/G5P6-4VAK)

28 Log-hub (2024) Inside the Formula 1 logistics race, https://log-hub.com/inside-the-formula-1-logistics-race/?trk=article-ssr-frontend-pulse_little-text-block (archived at https://perma.cc/4D5M-WWK9)

29 Mirza, Z (2024) Formula 1 'on track' to reach net-zero status by 2030 after emissions reduction: report, ESG Dive, www.esgdive.com/news/formula-1-on-track-to-reach-net-zero-status-by-2030-post-emissions-reduction/713628 (archived at https://perma.cc/C2FM-H2P8)

The role of automation and robotics in supply chain optimization

06

This chapter will discuss the role of automation and robotics in optimizing modern supply chain operations, including warehouse automation and transportation management.

CHAPTER AIMS

- Learn why automation and robotics are being adopted in supply chain management.
- Appreciate the benefits of automation and robotics in optimizing supply chain operations.
- Identify and build the case for automation and robotics.

'Do not follow where the path might lead. Go instead where there is no path, and leave a trail.'

MURIEL STRODE[1]

REAL-WORLD EXAMPLE The Very Group

The Very Group is a £2 billion revenue e-commerce retailer that offers consumers a range of products across fashion, electrical and home, with over

2,000 brands and own brand collections. It has over 4 million active customers and delivers 45 million items annually.[2] It evolved from a mail order catalogue business into an online business in 2009, and in 2020, it went live with an automated fulfilment centre called Skygate, based in the East Midlands in the UK. This location is desirable for warehouse operations due to it being in the 'golden triangle', which is an area in the UK where 90 per cent of the population can be reached within four hours.[3]

The project to build a new state-of-the-art warehouse involved the amalgamation of three distribution centres and was part of a strategic restructuring effort to reduce cost base, generate new sales and improve service to customers. One of the key elements of generating new sales comes from attracting shoppers in the evening, achieved with the ability to order and receive the next day. By implementing this new warehouse, the Very Group was able to extend the customer order cut-off time for next day from 7pm to 10pm. Also, it reduced the time from order to dispatch from four hours to 30 minutes, with it tracking a fastest time of 18 minutes during Black Friday.[4] The warehouse can handle 450,000 to 600,000 units in off-peak times and double that number per week in peak times.

Every box received is provided an ID, with 85 per cent of stock moved to a decant station where staff remove items and put them into totes, which are taken to the automated storage and retrieval system (ASRS). When an order is received, the tote is taken from storage to one of 14 'pick it easy' workstations, which eliminates the need for the 7 km walks per person to pick their daily demand. Once the order is picked from the tote, it then goes onto an evo-pocket system (a conveyor belt which operates from the ceiling of the warehouse) to move items to the packing floor area, which has storage for up to 120,000 pockets. From there, staff can pack items and then send them via conveyor belts to the dispatch sorter. Alternatively, if the product requires a box, it will go via the ASRS and a separate conveyor belt to be packed via an automated machine and then sent to the dispatch sorter (the aim of warehouse operations is to try to collate orders for the same customer into as few overall parcels as possible).

The dispatch sorter then directs orders to different carriers via a conveyor belt to reduce manual movement between 'goods out' and filling the trailer. Returns, which are a significant issue for e-commerce, were 16.9 per cent on average in 2024, according to the National Retail Federation, but could be as high as 30 per cent for some retailers.[5] Returns are treated like inbound items from suppliers except they are checked by staff at a station, given a quality rating and if acceptable, they can immediately have an 'available' status and go back to the storage system. The automation system is monitored from a control room (referred to as the 'flow' room) ensuring items move from A to B in the most efficient path. The operators in the control room will also work with engineers if issues arise to fix them quickly.[6]

In conclusion, the new warehouse has helped keep The Very Group competitive with benefits including the ability from 2021 onwards to receive and dispatch in 30 minutes compared to its previous four-hour timeframes,[7] dispatching over 210,000 orders on its busiest day of its financial year in 2023 and reducing operating costs in 2024 to the lowest level it had ever achieved.[8]

Optimizing the supply chain has long been a priority for companies as they strive to improve efficiency, reduce costs and exceed customers' expectations. With that, automation and robotics have established themselves as technologies that can greatly assist companies in addressing these priorities. From developing warehousing operations that use automation, to how products are transported through the supply chain by the introduction of autonomous vehicles, automation and robotics are assisting in streamlining operations, mitigating risks and shortening the time from order to delivery. The demand for faster service continues to be a differentiator for customers and who they choose to do business with. We have moved in a relatively short time from being content with receiving goods within a week to a few days, the next day and even the next few hours. With that comes increased financial costs but also potential opportunities that outweigh these. For example, a survey by Supply Chain Dive[9] found 80 per cent of respondents saw revenue increase with offering same day delivery, increases in revenue per order, net new sales, repeat purchases and increased conversion rates. Supply chain leaders must find new solutions to help solve the same-day delivery ask, and automation and robotics are part of the answer.

In the following sections, we will look at examples of popular technology solutions, the benefits derived from implementing them and how companies can identify and implement automation strategies to optimize their supply chain. It should be noted that I have not covered manufacturing as part of the scope of this chapter, primarily because robotics and automation in this area are well-established.

Key drivers of automation

As touched on previously, customer expectations have changed dramatically and continue to evolve. Customers expect to receive goods quickly, via a reliable delivery service and can return the goods, if necessary, with a similar experience. They, in many instances, want to have near real time tracking with flexible delivery windows and don't expect to have to pay much (if anything) for the delivery. For a company to deliver against these expectations requires a highly responsive and agile supply chain.[10] To have that type of supply chain in place requires a different mindset, which needs to focus on eliminating downtime between activities, introducing levels of automation to enable fast order processing, providing customers with an accurate view of products available to buy in near real time and orchestrating and executing against tight cut-off times.

The shortage of skilled labour and the rising cost of labour have become significant challenges for supply chain leaders. This initially led to many companies looking to offshore or outsource to cheaper markets to manufacture. Although this had a good effect on profit margins in the short term, in the long term, it has increased the skills and expectations of workers in those countries who now command higher salaries. This has resulted in higher costs for supply chain leaders to absorb globally, since they still need to get products from those locations to local markets. Automation provides companies with opportunities to mitigate some labour shortages, which has been especially pertinent in warehouse operations and transportation networks where labour is becoming increasingly scarce and expensive.[11]

Moving beyond labour costs, the pressure to reduce operational costs in general has never gone away and can be a barrier for companies if they don't get them under control. Automation offers a variety of ways to streamline operations and reduce waste and inefficiencies, as robots and automated systems aren't restricted by shifts.

The globalization and complexity of modern supply chains, coupled with managing them manually, is increasingly difficult to do. Therefore, companies need automation to manage sparse, large footprints with more accuracy.

Advancements in technology, specifically AI, ML and IoT, have accelerated the adoption of automation solutions, making them more accessible and at a cost that, as they become more popular, reduces significantly. These technologies are improving the capabilities of robots and automated systems, enabling them to perform more complex tasks with higher precision. When we think of the evolution of automation, we were excited by the ability to move pallets and then to move totes. However, the challenge has been how to use automation with items that require different ways to be picked up and handled.[12] Take an apple: you imagine it as round, but it can be oblate, oblong shaped, it can also taper at the bottom, can be ribbed or lobed. Therefore, a machine taught to pick up an apple needs to be able to determine how to do this potentially differently each time.

Advances in machine learning algorithms have given rise to autonomous vehicles and drones that can safely navigate complex environments. Similarly, advances in robots infused with AI are being deployed on warehouse floors, where they adapt to changing conditions and work collaboratively with human staff to execute tasks with precision and efficiency. They can pick and place different parts and materials and can help reduce automation costs up to 90 per cent.[13]

These drivers have helped the adoption of automation in companies beyond a select few who were the pioneers in this space such as Amazon (more on them later in the chapter).

Benefits of automation and robotics in supply chain

By introducing automation and robotics into supply chains, there are numerous benefits to be captured, including inventory management, warehousing, transportation and order fulfilment.

Improving inventory management and in turn, demand forecasting, are crucial first steps in optimizing performance of the supply chain, reducing costs and managing customer expectations. Automated systems, using technologies like RFID and IoT working in tandem with robotic inventory management systems, provide users and customers with accurate and up-to-date information on tracking and stock levels. AI and ML algorithms can help improve forecast signals by analysing many different datasets then adjust inventory levels dynamically, for example, balancing inventory across store estate or signalling to customers when inventory will be available. Automated systems can help companies proactively prevent stockouts or overstocking by providing near real time updates and assist in maintaining optimal stock levels. Operational costs can be improved using automation since the need to do manual checks and adjustments on inventory are reduced. The flow of data keeps the supply chain refreshed, updating inventory records, generating reports and providing users with information quickly to investigate issues rather than finding them during cycle counts – or not at all.

Warehouses are typically a significant cost of the supply chain and in most instances, are labour-intensive. Therefore, automation in these locations can lead to material improvements in efficiency, reduce errors and provide scalability for companies. The importance of the productivity metric lends itself to the use of automated systems – although the adoption has been slow, with estimates from 2024 suggesting that approximately 75 per cent of warehouses have no automation.[14] These systems can operate all day every day, handling repetitive tasks of the warehouse such as receiving, storage, picking, packing and sorting at speeds far beyond workers. Automation minimizes errors, reducing the role of humans in the process, especially when dealing with splitting pallets into cases / single items and making sure the right items move through the warehouse operation every time. This leads to improved order accuracy and customer satisfaction, which reduces returns and rework.

Optimization of warehouse space and layout is another consideration for companies when introducing robotics and automated systems. By having automated systems in place in particular parts of the warehouse, companies can increase throughput and use space more efficiently, such as expanding staging to get more goods out to customers in shorter time periods. Having spent time in over 200 warehouses in my career, you would be surprised that while they all follow similar processes, warehouse layouts and operations can vary dramatically even within the same company. One of the key reasons for this is many warehouses were set up for a different type of operation, for example, pallet in / out and servicing stores only. Now, they can service customers through a variety of methods and therefore need to split pallets, have warehouses with cross-docking capabilities (requiring goods to never be stored and moving from receiving to picking within a matter of hours) and a variety of picking options dependent on the product and route to market.

Modern warehouses are designed with automation systems considered within their business cases while warehouses built even up until 2020 may not have. I would like to think any company looking at warehousing today would be deploying a form of automation system within them or leaving space for them, aligned to growth plans.

Labour, whether from a shortage or cost perspective, is another factor to consider when looking at investments in automation systems and robotics. While initial investments in automation can be significant, over time, that investment can become fruitful when there is a reduction in manual labour required, reducing labour costs. Also, by using technology solutions in warehousing, companies can reduce their reliance on temporary or seasonal labour, which can fluctuate depending on demand, such as during retailers' peak seasons (Black Friday and Christmas).

Safety in warehouse operations is critical to a company and while automation and robotics won't fully eliminate the risk, it does help reduce workplace injuries by removing humans from dangerous and physically demanding tasks. Robotics can handle heavy lifting and reduce risks from repetitive tasks, improving worker safety.[15]

Furthermore, automated machines do not need breaks and can operate in the dark when no one is around. Depending on the type of automation and its purpose, they can replenish locations all day long. For example, an e-commerce retailer's orders come from an app or website. In many instances, purchases done via this route are completed in the evening since we have all been very busy at work with no time to stop and make purchases during the day (supposedly). If the operation is predominantly manual, then the following could be the process in the warehouse:

- Receive the order and check inventory in the network of warehouses.
- Confirm order and assign pick to a suitable warehouse dependent on a variety of factors (e.g. proximity to customer, inventory position, orders already received and allocated).
- Picks are collated in a warehouse management system and released (based on many different factors which could include volume of orders received, location in warehouse of inventory, where the customer address is).
- Picker locates the item and takes it to the packing area.
- Packer prints off label and packs item.
- Item is then moved to the staging area.
- Trailer comes onto the loading dock and the packaged item is scanned to the vehicle.

When you read these actions (and there may be additional steps) this all takes time. Although these could be done in an hour, the warehouse might have hundreds or thousands of orders, so by the time each separate order has been completed, it could take several hours for the tasks to be done. An important question to answer is when is the cut-off for placing orders and receiving next day? The customer might urgently require a dress for a social occasion the following night, therefore delighting or disappointing customers with the speed of delivery could have a significant impact on future business for the e-commerce retailer.

In this example, if we state the cut-off for receiving orders is 8 pm, then the warehouse operation has several hours to execute the above actions to deliver to the customer per expectations. However, if we now go back to the point of customers predominantly doing online shopping at night, the company will have potentially missed out on a sale because it could be that the customer only sits down to purchase at 10 pm. The pressing requirement of having the item delivered tomorrow for their event could lead to the customer happily paying to expediate the order, but if another e-commerce retailer has a similar item that is still available for delivery tomorrow when they go online at 10 pm, without the need to pay to expediate the order, this might lead to a sale and loyalty developing with the other company.

Moving beyond warehouses, transportation management is another adopter of automation and robotics that has had to innovate quickly. This is linked to the ever-reducing delivery times that customers now expect, coupled with the tension of having to achieve it in a sustainable manner. When it comes to automating transportation systems, the initial focus often lies in optimizing routes to minimize costs. This involves a range of strategies, from reducing fuel consumption and consolidating loads to maximizing trailer utilization and in some cases, replacing drivers with autonomous vehicles. However, with the advent of advanced network modelling and transportation management systems (TMS), companies can now analyse a vast array of variables – including topography, traffic patterns, weather conditions, trailer age, fill rates, temperature control and CO_2 emissions – to gain a deeper understanding of their cost-to-serve and identify opportunities for improvement. These systems also enable companies to reconfigure their networks in near real time, responding to changing conditions and customer demands while minimizing costs. By leveraging data-driven insights, businesses can make informed decisions about how to optimize their transportation operations, from adjusting routes and schedules to investing in new technologies.

Once considered the stuff of science fiction, self-driving vehicles and drones are now becoming increasingly mainstream, offering a more efficient and cost-effective means of delivery. As the technology continues to evolve and mature, we can expect to see widespread adoption across the transportation sector, transforming the way goods are moved and delivered.

It could be argued that improved delivery speed is probably one of the main priorities for customers and by using autonomous delivery vehicles and drones, companies can unlock better last-mile deliveries. These systems can reduce delivery times by bypassing traffic and operate 24 hours a day so they can be on the roads when they are less congested.

Human errors in transportation can result in delays and accidents but through automating transportation tasks, these errors can be minimized. Autonomous vehicles and robots follow optimized routes and they don't suffer from fatigue or distractions.

Fleet management can be improved using automation. Autonomous vehicles and drones can provide near real time data on their location, speed and delivery progress, which allows transport teams to track and adjust operations in the moment to make changes to keep operations going and to manage customer expectations in relation to deliveries.

Automation and robotic technologies in the supply chain

The use of automation has enabled companies to gain increased control and confidence of inventory management. Automations like RFID (see Chapter 4 for more details), use of supply chain technology systems (typically modules aligned to ERP) and robotic systems to scan shelves (using sensors and cameras) to detect missing or misplaced items, which then automatically update inventory records, are all examples of technology assisting users. For forecasting, AI and ML algorithms can analyse data sets to predict demand patterns and adjust inventory levels. There are several solutions to assist companies when used in conjunction that can help improve inventory management.

Warehouse automation and robotic technologies

Warehouse automation technologies have been around for a while now. The first automated storage and retrieval system (ASRS) was developed in the 1950s and commercialized in the 1960s.[16] They use conveyor belts, lifts and software to move items to and from storage locations at high speeds in an optimized space. They have gained popularity due to the growth of e-commerce and the direct-to-consumer business model[17] and are usually found where you have a high concentration of volume of pallets that need to be moved in and out constantly. Coca-Cola's Europacific Partners (CCEP) announced in late 2024 that they would build a 38-metre-tall system at Wakefield in the UK which was expected to be operational 30 months later.[18]

Many versions of warehouse automation technologies have been created to be able to handle different sizes of products (smaller and greater than pallet size) and in different forms. These technologies are becoming utilized by a variety of companies such as grocers and apparel retailers. Ocado, in its 'customer fulfilment centre' has a swarm of robotic vehicles that navigate a grid-like system to retrieve and transport grocery items to human packers. The bins, which are roughly the size of a large suitcase, move along a network of tracks and intersections, using a combination of navigation systems and sensors to avoid collisions and optimize their routes. At certain points on the grid are robotic arms that can pick and place items with precision, allowing for rapid and accurate order fulfilment. Once an order is received, the bins spring into action, retrieving items from various zones within the warehouse and transporting them to a central packing area. Employees then carefully assemble the orders, which are subsequently dispatched to customers.[19]

Automated guided vehicles (AGVs) are autonomous robots that are used to transport goods within warehouses, moving inventory between areas with limited human intervention. They reduce the need for human-controlled forklifts and manual labour, speeding up the material handling process. These AGVs have been used to follow staff as they conduct product picks, to moving order(s) from one location to another. They can be configured to move around following system-controlled routing and can safely engage with users throughout their journeys.

Drones, as mentioned previously in this chapter, are being used in warehouses and yards to help inventory checking and surveillance of the operation. Companies such as IKEA[20] and Walmart[21] are gaining benefits from them in their operations such as from reduced cycle counts, better yard management and knowing where inventory is (i.e. in racking or in a trailer). One of the main concerns with them that supply chain leaders must be aware of is the risk of operating in controlled environments restricted by height with humans working below.

REAL-WORLD EXAMPLE UPS

UPS, in 2024, invested in autonomous inventory tracking at one of its US warehouses where self-flying drones navigate the warehouse's narrow aisles, performing daily inventory checks, floor-level to top-rack scans and comprehensive audits. The system maintains accurate, near real time data on stock levels, enabling inventory visibility and control. At night, the drones capture high-resolution images in the dark, identifying discrepancies and conducting routine checks. The next day, the operations team receive a report showing any issues that can be quickly resolved. To date the work has led to reducing labour hours relating to inventory by 50 per cent (~8,000 hours).[22]

Robotic arms are being used for tasks including picking, sorting and packing products. These systems operate at speed with accuracy, minimizing human errors and improving throughput. From being used to de-palletize products onto conveyor belts, to being at the end of the picking line, and configuring and executing the building of a pallet, these solutions work well with predictable high volumes. Migros, the Swiss grocer, uses them in its operation to do the tasks mentioned above.[23]

There are many varieties of solutions being deployed in warehouse operations and I expect them to gain more popularity. As more companies assess their supply chains, they will look to modernize warehouses or outsource to logistics providers. The unlocking of significant benefits (e.g. productivity or lower cost) that do not require significant redesign of flows of goods makes them appealing.

Transportation automation and robotic technologies

Automation and robotics are playing a key role in transportation. We have been getting excited about autonomous vehicles for a while, especially when Elon Musk announced in 2013 that three years later, cars could be 90 per cent autonomous.[24] While Waymo has autonomous vehicles operating in three cities, the cost, safety concerns and which technology is most suitable (i.e. cameras versus lidar and radar) are still factors preventing the exploitation of this capability.[25]

Drones, beyond warehouses, are being used by delivery companies to aid in last-mile delivery. These are being used primarily to deliver small packages directly to customers. A successful example of it in action can be seen from Irish-based Manna,[26] which delivers a variety of products (e.g. food, medical supplies) to customers and is expanding its operations across Europe. In the US, Walmart partnered with DroneUp[27] to offer last-mile delivery services but is constrained in locations it can use them due to regulations. Logistics companies have been finding ways to use drones, with recent approval given to UPS to operate them beyond the visual line of sight in certain states.[28] In 2025, many larger organizations such as Tesco, conducted controlled pilots.[29]

Robotic systems are also being used to automate loading and unloading of freight from trucks and containers. For example, Joloda[30] can reduce the time for loading a container by 70 per cent to 80 per cent using automation versus a forklift. The use of these is most prevalent in port operations, where moving containers from dockside to the road is critical since they operate with limited capacity.

REAL-WORLD EXAMPLE Amazon

One company that has invested significantly in automation and robotics is Amazon. Its journey into automation and robotics began in earnest in 2012 with the acquisition of Kiva Systems, marking the beginning of enhanced efficiency, safety and productivity. Since then, Amazon has developed, produced and deployed more than 750,000 robots across its operations network, revolutionizing the way it handles fulfilment and delivery processes.[31]

The primary goal of incorporating robotics technology within Amazon's operations is to pair employees with the right technology to make their workday safer, easier and more productive, while also delivering packages to customers at speed. This vision has been realized through the development and deployment of a suite of advanced robotic systems that work in harmony to support package fulfilment and delivery. What is as impressive is its ability to deploy nearly identical technology in each of its fulfilment centres but change the layout as it learns what works when in operation, embracing the continuous improvement mindset.[32]

One of the latest milestones in Amazon's robotics journey is the launch of its next-generation, state-of-the-art fulfilment centre in the US in late 2024. This facility is equipped with the latest innovations in robotics, designed to support employees who package and deliver customer orders. The site uses eight different robotics systems that work together to streamline the fulfilment process.

For instance, it has a robotic system that stores and retrieves inventory for employees picking items for customer orders. It uses AI, robotics and computer vision systems to consolidate inventory and free up storage space, facilitating faster order transactions. Another robotic system transports pods of items to employees picking items for customer orders. It can lift approximately 550 kgs and navigate across large areas of the fulfilment centre.[33]

Another example is transporting large and bulky items, capable of lifting approximately 1,000 kgs. It is a robotic arm that detects, selects, handles and sorts individual products, placing them into totes for staff before packaging. Packaging automation systems create right-sized packages to fit an order's dimensions, reducing material usage.[34]

Amazon's first fully autonomous mobile robot moves carts with packages to the loading dock to be loaded onto trucks. It navigates freely throughout the site using sensors to detect and avoid objects, working in conjunction with other automation systems to streamline the outbound process.

Amazon's integration of robotics and AI has not only improved efficiency and productivity but also enhanced safety and job satisfaction for employees. An MIT study supports this claim: their research found that 60 per cent of employees who work with robotics and AI expect positive impacts on their productivity, job satisfaction and safety.[35]

In conclusion, Amazon's journey into automation and robotics has been marked by continuous innovation and integration of advanced technologies. The company's commitment to leveraging automation and robotics to support its workforce and enhance customer service has positioned it as a pioneer in the industry, setting new standards for efficiency and productivity in fulfilment and delivery operations.

What is obvious is that there are many automated systems and robotics available to companies to aid them in running their supply chains. The challenge is how to stitch these together to work in tandem with humans. I expect that there will always be a need for people within the physical movements through the supply chain, but the conundrum is to what level companies can automate without incurring expenditure, which makes it less appealing and flexible than using people in the process.

How to identify and build the case for automation and robotics

The case for automation and robotics investments in supply chain has become stronger, driven by the factors I have outlined in this chapter. However, like all advancements involving technology, investment requires careful planning, mobilization where possible through pilots to gain confidence and expertise in the solution and a business case that provides a return on investment. Furthermore, they should not be looked at in isolation but as part of an overarching strategy for the supply chain aligned to company goals. While some of the following steps may be obvious to you, many are given little attention: once decision makers are set on something, they are oblivious to other factors. There are plenty of stories about implementations that have gone astray, missed the mark on achieving the outcomes desired or been ripped out when teams couldn't make it work. For example, a consumer goods company invested more than $150 million on an automated warehouse which, when it went live, was not delivering against the business case. This investment was primarily driven by a decision based on inaccurate data, leading the company having to repurpose the warehouse to handle different channel requirements and having badly underutilized automation equipment sitting idle.[36]

Firstly, identifying opportunities for automation should begin by looking at the operations holistically with a focus on current and future pain points. An assessment of key metrics that inform how well the operation is currently handling demand and throughput should be done, as well as looking for and documenting inefficiencies, bottlenecks and where time and costs are being incurred through processes. The typical areas for automation and robotics as outlined in this chapter are primarily in inventory tracking, warehouse operations, order fulfilment and transportation, so I suggest focusing on assessing these.

The analysis of data and identification of repetitive tasks should be conducted to understand which of these are time consuming, prone to human error and have risk associated to them, particularly linked to safety. When reviewing data, caution should be taken to assess this with colleagues who are running the operation to ensure that it is based on reality. Checking the operation first hand is critical because it is at this stage that you need to know what you are saying is backed by facts. An example of this was the case involving KFC and DHL in 2018, where misunderstandings about the operation led to an unfavourable outcome for all parties. One party deemed the data it shared was sufficient for the other party to understand what was required and a warehouse was designed to meet that criteria, which led to the warehouse being unfit for purpose.[37] Although this example is not related to a technology implementation, it demonstrates the importance of understanding what the data is saying when it comes to supply chain operations. The data should be translated into people, process and technology requirements and played back and forth between parties.

The enjoyable part of the process comes next when evaluating solutions to understand their feasibility for your specific supply chain operations. I would like to make it clear here that not all automation solutions are suitable for every supply chain and to think otherwise is foolish. Consideration should be given to scalability and the integration capabilities aligned to your technology landscape – and please don't take the vendor's word for it! Ask to see solutions demonstrated with other customers of theirs. Attending conferences (like LogiMAT) can help educate decision makers on what is available and allows you to quickly understand the vendor landscape.

> Remember to ask about the availability of technology and engineers, since some systems are made to order and have considerable lead times.

Calculating the return on investment (ROI) is the main determinant of whether the automation/robotics initiative will proceed. The levers to examine here with supporting documentation should focus on cost savings, efficiency gains and other measurable benefits. Consideration for direct financial benefits (e.g. reduced labour cost, throughput) and indirect financial benefits (e.g. improved customer satisfaction, reduction in lost sales) should be documented with tested assumptions.

One of the key failures when producing a business case is the inability of the owner of the initiative to build a story that engages decision makers. The story should be explained in a simple consistent manner, address questions and concerns and build confidence in why the investment should proceed. Sometimes it can be down to timing or priorities, however, by the time the owner gets to writing the case, they should have engaged stakeholders and have the momentum to proceed with writing it (we will come back to this topic in detail in Chapter 9).

Conclusion

Automation and robotics have emerged as enablers to assist companies in optimizing their supply chains. Advancements in technology, which are ever occurring, can help companies get to grips with inventory management, streamline warehouse operations and optimize transportation management. These capabilities have demonstrated enhanced efficiency, reductions in operational cost that are sustainable and lead to improved customer satisfaction. To capture these benefits, companies must identify opportunities for automation and robotics, backed by a compelling business case.

As part of the overall supply chain modernization, companies need to adopt automation and robotics to aid them in agility, efficiency and resiliency because if anything is certain, the environment they operate in is only becoming more complex and competitors that can scale quicker are positioned to capture long-term growth.

Endnotes

1 Ratcliffe, S (2017) *Oxford Essential Quotations*, 5th edition, Oxford University Press

2 The Very Group (2025) About us, The Very Group, www.theverygroup.com/about-us (archived at https://perma.cc/VFX3-LMY3)

3 Morgan King (2021) Warehousing and the golden logistics triangle, Morgan King Blog, www.morganking.co.uk/blogposts/warehousing-and-the-golden-logistics-triangle (archived at https://perma.cc/SZN2-X8LT)

4 Wright, G (2021) Very extends next day delivery cut off time thanks to new automated tech, *Retail Gazette*, www.retailgazette.co.uk/blog/2021/08/very-extends-next-day-delivery-cut-off-time-thanks-to-new-automated-tech/#:~:text=Terms%20and%20Conditions-,Very%20extends%20next%20day%20delivery%20cut%20off%20time%20thanks%20to,for%20another%20strong%20Golden%20Quarter%2C (archived at https://perma.cc/ZH9W-Q8BZ)

5 Dopson, E (2025) Ecommerce returns: Average return rate and how to reduce it, Shopify, www.shopify.com/uk/enterprise/blog/ecommerce-returns (archived at https://perma.cc/S8JA-B2KU)

6 Knapp (2023) Knapp-Very Group, Knapp, www.youtube.com/watch?v=h6Agcdm5bf4#:~:text=Advanced%20automation%20technology%20from%20KNAPP%20is%20helping,new%20distribution%20centre%20in%20the%20East%20Midlands (archived at https://perma.cc/8JYL-2E6Y)

7 Mooney, M (2021) Orders processed in 30 minutes at The Very Group distribution centre, Logistics Manager, www.logisticsmanager.com/orders-processed-in-30-minutes-at-the-very-group-distribution-centre/#:~:text=An%20850%2C000ft2%20distribution%20facility,future%20growth%2C"%20he%20adds (archived at https://perma.cc/GS45-EKUX)

8 The Very Group (2024) The Very Group notes 'resilient' full-year performance, Insightdiy, www.insightdiy.co.uk/news/the-very-group-notes-resilient-fullyear-performance/14296.htm (archived at https://perma.cc/H4DL-G29B)

9 Supply Chain Dive studioID and Roadie (UPS Company) (2024) How the ROI of same-day delivery outweighs the cost, www.supplychaindive.com/spons/how-the-roi-of-same-day-delivery-outweighs-the-costs/709830 (archived at https://perma.cc/W3B2-NNCA)

10 Christopher, M (2024) Creating an agile supply chain, Cranfield School of Management, https://blog.som.cranfield.ac.uk/execdev/creating-an-agile-supply-chain#:~:text=To%20achieve%20this%20level%20of%20responsiveness%20requires,the%20business%20and%20across%20its%20supply%20chain.&text=Ultimately%20supply%20chain%20agility%20can%20only%20be,but%20rather%20as%20highly%20connected%20supply%20chains (archived at https://perma.cc/F6YX-UBY3)

11 Lee, I (2024) Addressing labour shortages in the logistics and transportation industry, *Global Trade Magazine*, www.globaltrademag.com/addressing-labor-shortages-in-the-logistics-and-transportation-industry (archived at https://perma.cc/ZV7S-22AV)

12 Del Rey, J (2022) Amazon's robots are getting closer to replacing human hands, Vox, www.vox.com/recode/2022/9/27/23373588/amazon-warehouse-robots-manipulation-picker-stower (archived at https://perma.cc/47NL-E2BU)

13 Schlauss, S (2025) 2025 Tech Trends Report, Future Today Strategy Group, https://ftsg.com/trends (archived at https://perma.cc/GZ7U-TL5F)

14 Straight, B (2024) Warehouse automation poised to rebound in 2025, Supply Chain Management Review, www.scmr.com/article/warehouse-automation-poised-to-rebound-in-2025 (archived at https://perma.cc/R9A4-PJZM)

15 Gihleb,R, Giuntella, O, Stella, L and Wang, T (2022) Industrial robots, workers' safety, and health, *Labour Economics*, 78, 102205, ISSN 0927-5371, https://doi.org/10.1016/j.labeco.2022.102205 (archived at https://perma.cc/D2HA-CD57)

16 Ashcroft, S (2022) Warehouse automation technology – from conveyors to robots, Supply Chain Digital, https://supplychaindigital.com/technology/warehouse-automation-technology-from-conveyors-to-robots (archived at https://perma.cc/E3E3-KVML)

17 Karnik, P (2023) Warehouse ASRS: Empowering e-commerce growth, Logistics Outlook, www.logisticsoutlook.com/logistics/warehouse-asrs-empowering-e-commerce-growth (archived at https://perma.cc/W289-B6E6)

18 Beeson, J (2024) Coca-Cola Europacific Partners pumps £42m into Wakefield factory site, *The Grocer*, www.thegrocer.co.uk/news/coca-cola-europacific-partners-pumps-42m-into-wakefield-factory-site/698233.article (archived at https://perma.cc/MF3B-E68U)

19 Ingham, L (2019) Inside Ocado: The automated warehouses powering technology's dark horse, Verdict_AI, https://verdict-ai.nridigital.com/verdict_ai_may19/ocado_automated_warehouse (archived at https://perma.cc/DTE3-TPKN)

20 Ingka (2024) IKEA is elevating human – AI powered drone collaboration, Ingka, www.ingka.com/newsroom/ikea-is-elevating-human-ai-powered-drone-collaboration (archived at https://perma.cc/T4NQ-F4RF)

21 Hargreaves, L (2024) Top 10: Digital warehousing solutions, Supply Chain Digital, https://supplychaindigital.com/articles/top-10-digital-warehousing-solutions (archived at https://perma.cc/R6U6-LJK7)

22 Verity (2024) Verity and UPS, www.verity.net/resources/verity-and-ups?utm_source=newsletter&utm_medium=email&utm_campaign=SCDOperations&utm_term=OperationsNewsletter (archived at https://perma.cc/4Z3K-6WU7)

23 Knapp (2019) KNAPP implements logistics solution at Migros Lucerne, www.knapp.com/en/knapp-implements-logistics-solution-at-migros-lucerne (archived at https://perma.cc/55CB-BJUH)

24 Charlton, A (2013) Tesla to produce self-driving car within three years, *International Business Times*, www.ibtimes.co.uk/tesla-self-driving-car-three-years-elon-507104 (archived at https://perma.cc/82A5-77VS)

25 Sants, A (2025) The stocks battling it out in the autonomous driving race, *Investors Chronicle*, www.investorschronicle.co.uk/content/cbb75c2f-5ae0-5e5e-b736-39b2282ceb16 (archived at https://perma.cc/Z5AT-UC57)

26 Heim, A (2024) Europe's drone-friendly regulations are helping Manna soar above US delivery rivals, Fortune, https://fortune.com/europe/2024/09/05/europes-drone-friendly-regulations-are-helping-manna-soar-above-u-s-delivery-rivals (archived at https://perma.cc/2PXM-VP4P)

27 Zukowski, D (2024) Drone delivery has last-mile potential. This CEO explains how it actually works, Smart Cities Dive, www.smartcitiesdive.com/news/drone-delivery-last-mile-droneup-ceo/735177 (archived at https://perma.cc/95PM-HL54)

28 Garland, M (2023) UPS gains critical exemption to operate drones beyond line of sight, Supply Chain Dive, www.supplychaindive.com/news/ups-flight-forward-faa-drones-beyond-visual-line-of-sight-matternet/693144 (archived at https://perma.cc/YL6R-YBZ2)

29 McCurdy, W (2021) Tesco to expand drone delivery trial, Retail Systems, www.retail-systems.com/rs/Tesco_To_Expand_Drone_Delivery_Trial.php (archived at https://perma.cc/5CG8-R3L3)

30 Joloda (2024) Automated loading and unloading solutions, Joloda Hydraroll, www.joloda.com/automation/solutions/#:~:text=Automated%20loading%20systems%20are%20built,it%20in%20under%205%20minutes (archived at https://perma.cc/SAH9-3ZP8)

31 McCann, K (2024) Amazon and Covariant partner to boost AI-powered warehouses, *AI Magazine*, https://aimagazine.com/articles/amazon-and-covariant-partner-to-boost-ai-powered-warehouses (archived at https://perma.cc/3YAW-W9BZ)

32 Mims, C (2021) *Arriving Today*, HarperCollins Publishers

33 Brown, A S (2022) Amazon's tiny robot drives do the heavy lifting, Amazon Science, www.amazon.science/latest-news/amazon-robotics-autonomous-drive-units-hercules-pegasus-xanthus-xbot (archived at https://perma.cc/GH46-3JHY)

34 Greenawalt, T (2025) Amazon has more than 750,000 robots that sort, lift, and carry packages – see them in action, Amazon, www.aboutamazon.com/news/operations/amazon-robotics-robots-fulfillment-center (archived at https://perma.cc/VBV8-537Q)

35 Armstrong, B, Chen, V K, Cuellar, A, Forsey-Smerek, A and Shah, J (2024) Automation from the worker's perspective: How can new technologies make jobs better? MIT, https://ipc.mit.edu/wp-content/uploads/2024/09/Automation_from_the_Worker_s_Perspective-30-Sep-pub.pdf (archived at https://perma.cc/Q9P3-HE5H)

36 Davies, A, Oca, A, Prieto, I and Sarilar, U (2023) Getting warehouse automation right, McKinsey & Company, www.mckinsey.com/capabilities/operations/our-insights/getting-warehouse-automation-right (archived at https://perma.cc/UH9B-QYX2)

37 Jones, M (2018) KFC and the supply chain fail, FCSI, www.fcsi.org/foodservice-consultant/eame/kfc-supply-chain-fail (archived at https://perma.cc/6AB6-8264)

Digital platforms for supply chain collaboration and integration 07

This chapter will discuss the role of digital platforms in enabling supply chain collaboration and integration, including cloud-based platforms and application programming interfaces (APIs).

CHAPTER AIMS

- Appreciate the importance of supply chain collaboration and integration in driving business outcomes.
- See how digital platforms enable near real time collaboration and integration.

'The cloud services companies of all sizes... The cloud is for everyone. The cloud is a democracy.'

MARC BENIOFF[1]

REAL-WORLD EXAMPLE Morrisons

Morrisons was founded in 1899 by William Morrison, who started the business as an egg and butter merchant. The first store opened in 1958 in Bradford, where its headquarters still reside today.[2] It is a grocery retailer with vertical integration operations (where it produces about half of its own products in-house) that went into private ownership in 2021 when Clayton, Dubilier & Rice acquired its issued share capital.[3] In 2022, Morrison then acquired McColl's wholesale business.[4] In 2025, it was the fifth largest supermarket in the UK with a market share of 8.6 per cent.[5] It operates nearly 500 stores and offers online shopping to nearly all UK residents, serving around 9 million customers weekly.[6] It also has a long-term supply relationship with Amazon[7] and has partnerships with Deliveroo and Just Eat as well.[8]

Morrisons operates its supply chain via seven distribution centres and makes over 1,000 deliveries to stores each day[9] with Eddie Stobart (a third-party logistics provider) responsible for transport.[10] From a technology perspective, its supply chain has been modernized and transformed over the last decade. In 2025, it has approximately 90 per cent of its systems connected through the cloud, targeting to make this 100 per cent over the next five years. It uses a variety of providers for cloud, including AWS.[11]

The supply chain technology requirements for retailers are slightly different to those of a manufacturing company, but what makes this more complex for Morrisons is the vertical approach by having its own manufacturing, plus wholesale operations and e-commerce.

The stores use a point of sale (PoS) system, which provides sales data in near-real time so they know at a holistic level what has been sold and therefore that can inform the upstream supply chain. This input informs its Forecast and Replenish (F&R) system;[12] its main purpose is to integrate demand forecasting with inventory replenishment to optimize product availability.[13] This data is shared with the warehouse management system (WMS), which triggers, via batch orders, what needs to be picked in the distribution centres (DCs) for delivery to stores. Signals are sent via their transport planning system, which analyzes the space on trailers and plans the make-up of the load to go to stores, which is then sent via electronic data interchange (EDI) to its third-party logistics provider to execute. Also, the ERP system will be recording all transactions from a financial perspective.

As this is going on, forecast versus actual sales data will be analysed by the supply chain planners to determine what products require making or ordering from manufacturing and other suppliers, with signals sent via systems (internally) or EDI (externally). While this is a simplistic way to describe the operation, it is very much an orchestration process enabled by technology.

From a store inventory perspective, in 2013, the operation was a lot different: staff would go round the store and determine what they needed and enter this into the system manually, which would then go to DCs. This might mean that if the store did the stock check at 5 pm, the DC would struggle to always push inventory to be picked and shipped to stores by the next day. Morrisons implemented digital order pads,[14] which store staff use to keep inventory records updated and they have evolved this solution since. For example, in 2024, it started testing and rolling out AI-powered cameras in stores to improve availability; the cameras provide real time updates on shelf data, which automatically directs staff to replenish if required.[15]

All this information helps to give the supply chain team confidence in what inventory they have available to sell via all channels. A key part of any successful supply chain operation is the underlying data. At Morrisons, its data is kept in a master data system which shares the data with other systems (this is a 'parent and child' relationship, as discussed further in Chapter 12) and is the single source of truth for each product. The system includes product code (SKU number), description, items per case, length, width and height of items, among other data. This is managed on Google Cloud, which Morrisons transitioned over to from a legacy data model in 2024; the change has helped it move from batch snapshots to near real time views of its supply chain.[16] One point that adds to this complexity for grocery retailers is the need for a supply chain that can be flexible and responsive with high throughput due to having a significant volume of perishable goods (e.g. fresh goods like fruit and vegetables with a short shelf life). Therefore, the importance of data and being able to act on it quickly is table stakes.

From an online perspective, Morrisons has implemented an order management system (OMS) that feeds to and takes from its website but also connects with Amazon to offer customers on that platform the same information. This requires sharing of data across systems via the cloud and again highlights the importance of availability and knowing what it can offer versus inventory. Following the OMS implementation, Morrisons has evolved its supply chain systems following a strategy that embraces modernization and relies on having a strong internal team of technology professionals, complemented with third-party integrators who have specific skill sets for the solutions being implemented. This approach keeps the control firmly with the Morrisons team and it is prudent with how it goes about implementing technology within its supply chain to ensure this.

Morrisons follows a rolling three-year strategy, which is built and iterated annually in collaboration between supply chain and IT. Importance is placed on developing a technology roadmap that aligns with its business objectives, requiring clear problem statements and business buy-in for funding. The leadership team has a willingness to embrace technology investment and place significance on the necessity of evaluating technology providers' solutions for compatibility with existing systems.

By ensuring availability and coupling this with a 'secure by design' approach, its criteria for decision-making are firmly focused on improvements that will have a positive impact on its promise to customers. It embraces digital platforms to improve its supply chain operations from farm to fork, where Morrisons and its partners are using near real time data, collaborating to exceed customer expectations. It is a journey that often works at a considered pace whereby Morrisons endeavours to not rush and is content for implementations to take longer, so risks to current operations are minimized. That might sound like common sense; however, many other companies take a more aggressive approach. Countless benefits have been achieved by the way Morrisons has modernized its supply chain performance, including efficiencies of £1 million in warehouse operations.[17]

The advancements in technology and the amount of data at our fingertips, coupled with our need to know what is going on globally, have provided us the ability to collaborate seamlessly across the company's global footprint and extended supply chain much more effectively than ever before. Therefore, we have had the opportunity to gain more control and understanding of how it is performing in near real time. Our heritage of siloed information within the supply chain is outdated and no longer fit for purpose to keep pace with the speed, complexity and volume of global business.[18] As a result, the need for technology continues to be at the forefront of investment decisions at the board level, as digital platforms that include cloud-based systems enable collaboration and integration between a host of different tools, including enterprise resource planning (ERP), supply chain-specific tools, application programming interfaces (APIs) and other digital ecosystems. These platforms help companies co-ordinate activities and provide the infrastructure to connect and share data beyond the company's four walls, leading to greater transparency, efficiency and flexibility.

In this chapter, we will review the role of these digital platforms as enablers of collaboration among different systems and their integration into the supply chain. We will examine the importance of collaboration and outline some of the main digital platforms that are transforming the supply chain and how companies can use them to drive business success.

In my experience, companies usually don't choose to fully align with cloud products when managing their supply chains. They typically pick and choose which supply chain systems are most important for their company to use across their company footprint internally and which ones to provide access for partners to engage with. Also, return on investment is crucial since cloud services incur operational expenditure, so they need to be justified regularly, ensuring that they are adding value from tangible as well as intangible perspectives. Tangible value comes from not having to invest so much in hardware, on-site infrastructure and reducing maintenance costs. Intangible value is gained from being able to collaborate through seamless sharing of data regardless of location, as well as the ability to be flexible on resourcing throughout different periods of time, for example.[19]

The need for cloud-based platforms and APIs

The importance of collaboration and integration for supply chain success should not be underestimated. It is the process by which stakeholders across the supply chain (i.e. suppliers, manufacturers, logistics providers and customers) work together to achieve a shared goal, improve processes and optimize performance of the overall supply chain, not to the detriment of any party.

Collaboration

Effective collaboration helps eliminate silos, promoting a togetherness that is more dynamic and responsive to the supply chain. This has become even more essential for companies that are having to operate in volatile, uncertain, complex and ambiguous (VUCA) environments.[20]

Collaboration can at times sound like a buzzword. However, the key benefits of it when looking through a supply chain lens include shared information, decision-making that enables joint success, shared risk and continuous improvement and innovation.

Information sharing

Information sharing enables movement beyond your own part of the supply chain. Critical information sharing can be in relation to inventory (at various stages); demand forecasts and anomalies that are explainable; production schedules and priorities; tracking of shipments and where items are (or expected to be in time) in the supply chain. The importance of having timely and accurate information that is shared between companies helps reduce the chances of errors, misunderstandings and stockouts or overstocking, which can lead to unnecessary costs as well as significant waste. Significant steps are being taken by many leading organizations to improve their position on waste and sustainability in general. For example, Procter & Gamble (P&G) partnered with TerraCycle to create the 'loop' platform. It is a circular delivery system that lets customers purchase from P&G using reusable containers, which are collected, cleaned and refilled using a shared digital platform for customers to engage with.[21]

Decision-making

Decision-making made jointly between companies facilitates better outcomes because it incorporates the knowledge and insights of multiple parties. This can be important when making trade-offs based on constraints and coming to decisions on pricing, delivery schedules, production capacity and inventory.

In complex global supply chains, exogenous shocks can disrupt operations, but by having strong collaboration with partners, companies can distribute risk effectively and put in place contingency plans that can be executed at pace. During the Covid pandemic, we saw many examples of this play out where the chain had to bend and bounce more than usual. Tesco worked with its toilet roll suppliers to simplify having 33 variants to 10, and drive volume, therefore allowing suppliers to run longer production runs and reduce the chance of stockouts.[22] This worked at speed due to the ability to collaborate within systems, as well as each company's teams being able to see and respond to sales in near real time. Unilever was similarly able to make tactical decisions with the visibility it had through its systems being on cloud-based platforms to respond and engage with suppliers and customers.[23]

Continuous improvement and innovation

Collaboration encourages and builds a mindset where partners think beyond their own operation and consider the value chain and share best practices, technology advancements and process improvements. Working through these considerations together provides companies with the ability to develop solutions to overcome common challenges and share openly where one company's activities might lead to a positive outcome for them but be detrimental to another company, resulting in a high cost to deliver across the wider value chain. Examples of good results from collaboration can include helping to reduce lead times, improving quality of products and optimizing transportation routes.

Integration

People and processes are critical to operating supply chains but collaboration via connecting systems and integrating the supply chain is rapidly becoming as important, as this supports the workflows between different companies that make up the extended supply chain. Integration is key to this, which is focused on automating and streamlining the sharing of data and processes so that there is a seamless flow of information that supports activities between different functions in the wider supply chain ecosystem. For example, Maersk offer their customers the ability to integrate their warehouse management system (WMS) with customer systems to exchange information and provide visibility.[24] Maersk has the view that when collaborations result in end-to-end supply chain integration, they are most valuable.

Integration between companies and systems allows the sharing of data in near real time. With this visibility, companies can monitor the movement of products, inventory levels and order status from nth tier suppliers to customers. This helps reduce inefficiencies and improves decision-making with having the latest information to aid you at your fingertips. Integration also enables reducing latency (the delay before transferring data) between suppliers, manufacturers, retailers and customers. Without the integration, any supply issue or constraint requires multiple MRP runs in different ERP systems and multiple planning cycles that are not synchronized. This can result in a delay of several weeks before information is relayed to appropriate manufacturing sites or the customer learns about an issue.

Having integrated systems with automation also reduces (if not eliminates) the requirements for manual data entry and reconciliation, reducing errors and accelerating processes. For example, the duplication of work, such as having to re-enter data into multiple systems across companies, is minimized. This leads to reduced lead times, efficient use of resources, lower risk of stockouts or overstocking, faster order fulfilment and cost reductions.

Integration and visibility link to agility and responsiveness in the supply chain. Agility comes from being able to respond to changes in demand, act on supply disruptions quicker and adapt to market conditions through near real time updates and co-ordination.

Cloud-based systems and APIs are the backbone of modern supply chain integration and collaboration and have revolutionized how supply chains operate today. These platforms provide the infrastructure for seamless communication, data exchange and process automation across the companies that make up the end-to-end supply chain. We will now explore how they enable this collaboration and integration.

It is first appropriate to outline the three most common configurations of cloud-based platforms that support supply chains:

- **Software-as-a-Service (SaaS)** tools are hosted and distributed by third-party providers, eliminating the need for infrastructure changes with automatic software updates and maintenance in the background, ensuring seamless access for companies.
- **Infrastructure-as-a-Service (IaaS)** is a cloud-based model for data management and analysis, including logistics and supply chain technologies such as IoT sensors, virtual machines, networking technology and energy-management infrastructure for comprehensive support.
- **Platform-as-a-Service (PaaS)** is a stable, cloud-based development environment for testing cloud services and ensuring cross-compatibility with existing products and APIs. This is suited for supply chain management, where partners use diverse products from various vendors and require seamless integration and testing.

There are many cloud-based platforms in the market today that offer one or more of the above configurations. In 2024, Amazon Web Services (AWS), Microsoft Azure, Google, Alibaba and IBM made up over 70 per cent of the market.[25] In a supply chain context, there are also many other SaaS providers who offer solutions across the source, plan, make, move and return spheres, including companies such as Oracle and SAP.

Whatever workflows companies choose to run, their supply chains with cloud environments can exist on private or public clouds with a combination of these known as a hybrid cloud. Figure 7.1 shows a cloud-based platform.

Figure 7.1 A cloud-based platform

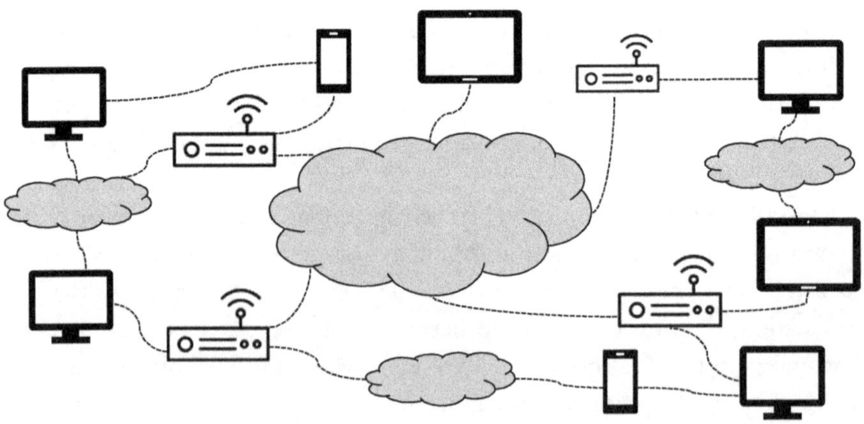

Now that we have covered cloud-based platforms, we have to find a way to connect systems across it to communicate through sharing data, features and functionality, which is where application programming interface (API) is required.[26] These APIs have a set of rules / protocols that share only what is required so they help secure systems as well within the cloud environment.

Below are the most widely used APIs for connecting supply chain systems (also see Figure 7.2). Their use is dependent on the industry, system and what data the company is wanting to share with other parties. While most of us don't need to know the ins and outs of them, it is helpful to understand the terminology.

- **RESTful APIs** are web APIs based on representational state of resource (REST) architecture. These APIs use hypertext transfer protocol (HTTP) methods (GET, POST, PUT, DELETE) to interact with resources. Examples of these include JSON-based APIs, which exchange data such as product information, inventory levels and shipping status. XML-based APIs, which provide structured and standardized ways of exchanging data are popular as well and are used regularly for proof of deliveries.[27]
- **EDI (electronic data interchange) APIs** are very popular in supply chains and are used to exchange documents such as purchase orders, invoices and shipping notices. They were first introduced in the 1960s[28] and were widely adopted in the 1980s, with Walmart providing access to point of sale (PoS) data and inventory information to thousands of suppliers via EDI by the early 1990s.[29] In the 2010s, they developed into being cloud-based, which is what companies use today.
- **Webhooks APIs** are popular to support sending near real time notifications between supply chain systems such as shipping updates and inventory changes.[30]
- **GraphQL APIs** are query-based interfaces for accessing data in supply chain systems such as product information and inventory levels.[31]
- **API-based integration platforms** provide pre-built connectors for more standard popular supply chain systems. Two examples of this type of platform are MuleSoft and Boomi.[32]
- **Industry-specific APIs** such as GSI, which is widely used in retail and consumer goods sectors to exchange information, and ATA, which is popular for exchanging logistics data in the transportation sector.[33]

As mentioned previously, one of the more popular uses of APIs in relation to EDI is the electronic exchange of business documents between companies using a secure and reliable communication protocol. The process is as follows:

1. Document creation – a document, such as a purchase order, is created electronically.
2. Translation – the document is translated into a standardized format, such as X12 or EDIFACT. The X12 standard is widely used in North America while EDIFACT is widely used in Europe and other parts of the world.
3. Transmission – the document is transmitted to another company's system (e.g. supplier / customer) using a secure communication protocol, such as file transfer protocol (FTP) or applicability statement 2 (AS2).
4. Receipt – the other company receives the document and processes it electronically.

Benefits of cloud-based platforms and APIs

The use of cloud-based platforms in the supply chain has grown exponentially since around 2015, growing eightfold in two years due to improved access to data and adoption of technologies such as IoT.[34] Cloud-based platforms provide companies with flexibility, scalability and reliability to collaborate and integrate across their (and other) parties in the supply chain. Unlike traditional on-premises solutions, cloud platforms provide near real time data access, which has become imperative for companies managing global supply chain operations that involve multiple parties located across the globe.

These advantages of near real time data sharing cannot be overstated. Data that is immediately accessible to all parties is fast becoming a given. This data provides transparency and assists decision-making, which improves inventory management and response times.

Cloud-based platforms consolidate data from various systems across the different activities of supply chain. Using this centralization helps to minimize

Figure 7.2 APIs

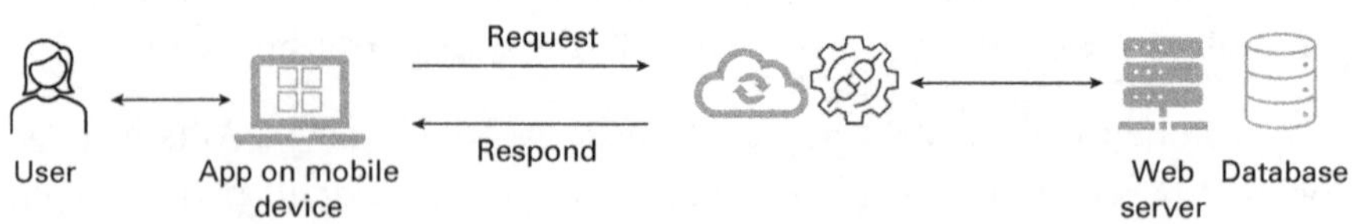

data silos, making it easier for users to analyse data leading to more accurate insights. The Unilever example in Chapter 5 covered this.

These platforms support businesses to scale their operations as needed without having to invest in significant amounts of infrastructure. This provides flexibility to companies with global supply chains that need to accommodate geographic expansion or contraction with fluctuations in demand, for example.

Through leveraging cloud computing, companies can allocate capital expenditure on infrastructure and only incur costs for the resources they use. This makes it much more affordable for different companies in the supply chain when moving upstream to suppliers of suppliers. Now, there is a somewhat level field between small, medium-sized and large enterprises.

Improved collaboration is facilitated by cloud platforms and the ability to communicate and collaborate between teams, suppliers, distributors and customers. This is achieved through integrated messaging systems, collaborative planning tools and shared dashboards, for example, which makes it easier for all parties to work together on a single platform. Ivalua[35] offers a supplier relationship management (SRM) solution that provides suppliers with a platform that they can self-service with on a variety tasks, from submitting bids to managing orders. SAP transportation management capability is another example of that, where a company can offer shipments details and freight providers who are on their system can quote against it to secure the shipment.[36]

The rise and popularity of cloud platforms has occurred with APIs too, as more and more companies find ways to stitch data across their enterprise and beyond. The open APIs market was valued over $4.5 billion in 2024 and is expected to grow significantly by 2033 to over $31 billion.[37] This is being driven by the emergence and embracement of new technologies such as Generative AI to share and use more data to aid companies.[38]

APIs enable the integration of different systems both within and outside of a company's supply chain. They allow disparate systems such as ERP, TMS, WMS and supplier platforms to communicate and exchange data seamlessly. This streamlined data exchange has been a significant advancement over the last 50 years, as now APIs enable near real time data exchange between systems. For example, when a customer places an order, an API can send the order details to the company, but potentially also key suppliers linked to specific components. It can also allocate and update the inventory system, automatically instructing the warehouse team via the system to generate the next pick list to fulfil order process. This helps to eliminate manual data entry, to avoid disappointing customers when inventory says it is available but isn't, which reduces the risk of errors and reduces the time across supply chain activities.

APIs allow companies to customize their supply chain systems and integrate third-party applications where appropriate. This provides flexibility and customization, which helps companies to adapt to evolving business requirements, such as having to integrate a new logistics provider or adding a new tool.

APIs also minimize system fragmentation, where traditionally a company's supply chain had a variety of systems operating in silos on different versions with limited (if any) ability to communicate with each other. APIs eliminate these silos and expand the capacity for companies to collaborate with each other by enabling systems to share data and create an integrated and cohesive supply chain. Also, it is easier to integrate new systems and platforms as companies become more complex. APIs allow companies to extend their digital supply chain networks at speed and efficiently, reducing the time to onboard or offboard partners.

From an innovation perspective, APIs assist companies in being able to rapidly adopt new technologies and services. With APIs, companies can integrate tools without having to overhaul their current supply chain system. This becomes critical as new technology arrives, and companies need to respond to new needs.

Considerations when implementing digital platforms

When implementing digital platforms into your company's supply chain, collaboration and integration requires considered planning, execution and ongoing optimization. Below are some of the best practices to consider.

Before ploughing into adopting digital platforms, companies should be clear on the objectives they want to achieve for collaboration and integration. It could be one or many, such as to improve inventory management across the supply chain, reduce lead times or enhance customer service. Being specifically upfront about their goals should help ensure that the chosen platforms align with the supply chain strategy.

For any initiative, it is vital that the team secures buy-in from all relevant interested parties. These should include senior leadership, IT teams, users and external partners. While some companies have the clout to demand that suppliers use their platform (i.e. due to their size and dominance in a market, for example), for most it is about influencing them on the benefits of being joined through systems. The tools must be user-friendly and all users must be able to understand the value of the platform to result in successful adoption. Even with these measures in place, it can be still a challenge. The procurement team of your customer in the supply chain may focus on cost savings within their process, thus selecting the cheapest supplier without considering the capability to share information through integration. Also, there can be conflict when a supplier is delivering to multiple competitive manufacturers, for example, where there is not transparency on constraints and allocations. Finally, customers can be reluctant to share their promotion calendars with manufacturers, afraid that the information may be leaked to their competitors and they will react with a similar promotion before their event to load the consumers.

The selection of the right platform can be time-consuming as there isn't a one-size-fits-all solution for digital platforms. Companies must evaluate different platforms through a variety of lenses, including their unique needs, scalability, integration capabilities, budget and ongoing costs. Consider cloud platforms versus on-premises platforms. The latter should be best avoided, if possible, due to lack of flexibility and possible risks to API support and vendor support services if it were to go down.

With the need for interconnectivity across the supply chain, data security becomes a critical companion. As we have seen, when a breach occurs, it can have a significant impact on a company and shows how fragile it can become. Morrisons and Starbucks had one of their SaaS providers hacked in November 2024. For Morrisons, this meant an outage to its warehouse management system for fresh food and produce, which meant it had challenges with inventory and replenishing stores.[39]

Once the platform is embedded, continuous monitoring of its use and optimization is important. Tracking key performance indicators (KPIs) such as order accuracy, supplier performance and inventory turnover will show how you are progressing or where optimization is required. This should be coupled with regularly assessing the effectiveness of the platform and making interventions where needed. Also, considering sequencing of additional capabilities and how the platform needs to evolve will keep it relevant and keep users engaged.

Establishing and integrating a digital platform in the supply chain is a significant undertaking. Companies should reflect on it as journey, following a phased approach whereby starting small and getting quick wins is crucial to building confidence and expanding out following a rigorous process. First, get the company's internal supply chain connected – this should be the priority post internal approval. Then move beyond, across the entire supply chain and connect to the customer side.

Conclusion

To enable supply chain collaboration across global footprints and with parties upstream and downstream requires companies to collaborate and integrate their processes. Digital platforms are the nucleus to enable this, providing the tools to streamline processes, enhance agility and sustainably reduce costs across the entire supply chain. Cloud-based solutions and APIs facilitate data sharing in near real time, helping companies respond to changes in their markets and customer demands faster. As companies look for a competitive edge, the adoption of digital platforms assists them in embracing an increasingly more digital and interconnected world by helping to improve supply chain collaboration and integration. Making considered decisions on the platforms that fit with the desired outcomes, having alignment among stakeholders and being on the journey of continuously optimizing processes, should unlock significant benefits for a company's supply chain and partners, such as improved efficiency, customer satisfaction and profitability.

Endnotes

1 Benioff, M and Adler, C (2009) *Behind the Cloud: The Untold Story of How Salesforce.com Went from Idea to Billion-Dollar Company-and Revolutionized an Industry*, Wiley-Blackwell

2 Himelfield, D (2021) A history of Morrisons – from egg stand to Britain's fourth largest supermarket chain as takeover bid accepted, Yorkshire Live, www.examinerlive.co.uk/whats-on/shopping/gallery/history-morrisons-egg-stand-britains-20963134 (archived at https://perma.cc/WTQ5-PN3N)

3 Eley, J (2021) CD&R triumphs in £10bn Morrisons battle, *Financial Times*, www.ft.com/content/4d990b0f-2728-4126-adee-d4111aae7a45 (archived at https://perma.cc/C7P8-CS8E)

4 Reuters (2022) UK's Morrisons clinches deal for convenience chain McColl's, Reuters, https://www.reuters.com/business/morrisons-clinches-deal-uk-retailer-mccolls-sky-news-2022-05-09/#:~:text=In%20a%20deal%20structured%20through%20a%20so,deal%20with%20McColl's%2C%20will%20take%20on%20all (archived at https://perma.cc/UUL5-LH57)

5 Louis, Y (2025) Lidl and Ocado thrive as Asda's market share slumps, *Grocery Gazette*, www.grocerygazette.co.uk/2025/02/04/lidl-ocado-thrive-asdas-slump (archived at https://perma.cc/3MX5-9ELD)

6 Morrisons Corporate (2025) Company history, www.morrisons-corporate.com/about-us/company-history (archived at https://perma.cc/Y3HK-GPJ5)

7 Deans, J (2016) Amazon enters fresh food market with Morrisons deal, *The Guardian*, www.theguardian.com/business/2016/feb/29/amazon-enters-fresh-food-market-with-morrisons-deal (archived at https://perma.cc/QDP6-8FVN)

8 Radojev, H (2024) Morrisons expands Just Eat partnership to include supermarkets nationwide, *Retail Week*, www.retail-week.com/grocery/morrisons-expands-just-eat-partnership-to-include-supermarkets-nationwide/7046740.article (archived at https://perma.cc/UKM4-TVNT)

9 Morrisons (2025) Logistics, Morrisons, https://www.morrisons.jobs/our-teams/logistics#:~:text=With%20seven%20sites%20across%20the,products%20at%20the%20right%20time (archived at https://perma.cc/8PT9-F573)

10 MotorTransport (2023) Wincanton woes mount as Morrisons switches £70m transport contract to Stobart, MotorTransport, https://motortransport.co.uk/wincanton-woes-mount-as-morrisons-switches-70m-transport-contract-to-stobart/16859.article (archived at https://perma.cc/92EZ-CL9L)

11 Mastek (2024) Driving retail digital transformation: Morrisons & Mastek, Mastek, www.mastek.com/resources/driving-retail-digital-transformation-morrisons/#:~:text=The%20ongoing%20partnership%20with%20Mastek%20encourages%20innovation%2C,driving%20business%20growth%20and%20improve%20operational%20efficiency.&text=By%20working%20with%20Mastek%2C%20Morrisons%20have%20been,wider%20customer%20base%20and%20increase%20brand%20reachability (archived at https://perma.cc/WMY9-AKFB)

12 Yates-Roberts, E (2018) Morrisons implements Blue Yonder's AI stock replenishment technology, Technology Record, https://www.technologyrecord.com/article/morrisons-implements-blue-yonders-ai-stock-replenishment-technology#:~:text=UK%20supermarket%20chain%20Morrisons%20has%20implemented%20Blue,right%20level%20of%20stock%20and%20reduce%20waste.&text=The%20Blue%20Yonder%20Demand%20Forecast%20&%20Replenishment,two%2D%20to%20three%2Dday%20decrease%20in%20stockholding%20in%2Dstore (archived at https://perma.cc/R7VL-VK97)

13 Rode, J (2023) Forecasting and replenishment: How hypermarket Kaufland nailed it, The Future Of Commerce, www.the-future-of-commerce.com/2023/06/26/forecasting-and-replenishment-kaufland (archived at https://perma.cc/8TBR-RLXF)

14 Thoughtworks (2013) Morrisons starts rollout of electronic order pad, Thoughtworks, www.thoughtworks.com/en-gb/about-us/news/2013/morrisons-completes-rollout-of-electronic-order-pad (archived at https://perma.cc/TDJ5-F8KZ)

15 Hill, E (2024) Morrisons overhaul: One year in, what changes has CEO Rami Baitiéh made? *Retail Gazette*, www.retailgazette.co.uk/blog/2024/11/morrisons-rami-baitieh-changes (archived at https://perma.cc/F2NJ-X7R3)

16 Striim Team (2024) Morrisons updates data infrastructure to drive real-time insights and improve customer experience, Striim, www.striim.com/blog/morrisons-data-infastructure (archived at https://perma.cc/2TM4-B2HW)

17 Aptean (2024) WMS software creates £1M annual Morrisons cost savings, Aptean, www.principallogisticstechnologies.com/warehouse-management-software-morrisons-supermarket-case-study (archived at https://perma.cc/U874-7R3R)

18 Business Reporter Writer (2025) Unchained: why digital platforms in supply chain are a game changer, *Business Reporter*, www.business-reporter.co.uk/management/unchained-why-digital-platforms-in-supply-chain-are-a-game-changer (archived at https://perma.cc/F7YH-TLED)

19 Patterson, D (2021) Economics of cloud computing: Driving cost efficiencies to fuel growth, *Business Tech Weekly*, www.businesstechweekly.com/operational-efficiency/cloud-computing/economics-of-cloud-computing/#:~:text=The%20general%20objective%20is%20to,and%20increased%20speed%20to%20market (archived at https://perma.cc/5AE6-CETV)

20 Metersky, J (2024) In the VUCA era, it's all about trade-offs, SupplyChainBrain, www.supplychainbrain.com/blogs/1-think-tank/post/40348-in-the-vuca-era-its-all-about-trade-offs (archived at https://perma.cc/WQ88-49BD)

21 Cotter, N (2024) 10 (Real World) Supply chain partnership examples, Growann, www.growann.com/post/supply-chain-partnership-examples#4-retailers-and-e-commerce-platforms-optimize-fulfillment (archived at https://perma.cc/3UXB-Q5JC)

22 Perry, M (2020) Five companies that reset their supply chains, Raconteur, www.raconteur.net/supply-chain/companies-supply-change (archived at https://perma.cc/WDZ5-VK9Z)

23 Berthiaume, D (2021) CSA Exclusive: Unilever jump starts slow-moving goods supply chain, Chainstoreage, https://chainstoreage.com/csa-exclusive-unilever-jump-starts-slow-moving-goods-supply-chain (archived at https://perma.cc/RV96-DTG2)

24 Maersk (2025) What sort of visibility can Maersk provide me when goods are under your custody? Maersk, www.maersk.com/support/faqs/sort-of-visibility-can-maersk-provide-me (archived at https://perma.cc/E69V-3ULN)

25 Slingerland, C (2025) 13 top cloud service providers globally in 2025, Cloudzero, www.cloudzero.com/blog/cloud-service-providers (archived at https://perma.cc/W3MS-MEYX)

26 Goodwin, M (2024) What is an API (application programming interface)? IBM, www.ibm.com/think/topics/api (archived at https://perma.cc/G7KD-GCAR)

27 Bigelow, S.J (2024) RESTful API, TechTarget, www.techtarget.com/searchapparchitecture/definition/RESTful-API (archived at https://perma.cc/K5NZ-L345)

28 Bate, A (2024) EDI vs API – The evolution of EDI and its role in modern business: From paper to pixels, Intelligent CIO, www.intelligentcio.com/eu/2024/09/05/edi-vs-api-the-evolution-of-edi-and-its-role-in-modern-business-from-paper-to-pixels (archived at https://perma.cc/Q5KD-7ULK)

29 Meier, S and Oberholzer-Gee, F (2015) Wal-Mart in search of renewed growth, Columbia Business School, https://g.co/kgs/o7zr5DP (archived at https://perma.cc/P5Q7-QWCC)

30 Red Hat (2024) What is a webhook? Red Hat, www.redhat.com/en/topics/automation/what-is-a-webhook (archived at https://perma.cc/859S-D7B2)

31 China, C R and Jackson, G (2024) What is GraphQL? IBM, www.ibm.com/think/topics/graphql (archived at https://perma.cc/6L56-U5VS)

32 Davies, A (2024) Top 5 API integration tools for your business: An in-depth guide, OpenLegacy, www.openlegacy.com/blog/api-integration-platform (archived at https://perma.cc/DJ5S-Q23K)

33 Mohite, J (2024) Securing APIs while navigating today's booming API economy, Akamai, www.akamai.com/blog/security/securing-apis-while-navigating-todays-booming-api-economy (archived at https://perma.cc/H2JA-3UXM)

34 Kochar, A (2019) Digital supply chain transformation: Taking the first step, Supply Chain Xchange, www.thescxchange.com/articles/2061-digital-supply-chain-transformation-taking-the-first-step#:~:text=The%20adoption%20of%20digital%20technologies%20has%20surged,could%20more%20than%20double%20again%20by%202021 (archived at https://perma.cc/X3CG-R7LN)

35 McMillan, A (2024) Top 10: SRM platforms, *Procurement Magazine*, https://procurementmag.com/top10/top-10-srm-platforms-for-procurement (archived at https://perma.cc/67AX-H88F)

36 SAP (2025) What is SAP transportation management? SAP, www.sap.com/uk/products/scm/transportation-logistics.html (archived at https://perma.cc/M5GY-LJY9)

37 Rai, R (2024) Open API market size, share, growth & trends chart by 2033, Straits Research, https://straitsresearch.com/report/open-api-market#:~:text=High%20demand%20from%20developers%20to%20spur%20market%20opportunities&text=According%20to%20the%20same%20poll,driving%20demand%20in%20various%20sectors (archived at https://perma.cc/TD9S-DXUA)

38 Gartner (2024) Gartner predicts more than 30% of the increase in demand for APIs will come from AI and tools using large language models by 2026, Gartner, www.gartner.com/en/newsroom/press-releases/2024-03-20-gartner-predicts-more-than-30-percent-of-the-increase-in-demand-for-apis-will-come-from-ai-and-tools-using-llms-by-2026 (archived at https://perma.cc/4ZUD-6KLD)

39 Jones, D (2024) Morrisons recovers warehouse systems following attack on Blue Yonder, Cyber Security Dive, www.cybersecuritydive.com/news/morrisons-recovers-attack-blue-yonder/734863 (archived at https://perma.cc/F4FD-9V8S)

The impact of supply chain technology on sustainability and social responsibility

08

This chapter will discuss the impact of supply chain technology on sustainability and social responsibility, showing how companies are trying to address these topics with varied levels of success.

> **CHAPTER AIMS**
>
> - Understand the role of supply chain technology in helping to address sustainability and social responsibility.
> - Recognize the importance of environmental and social considerations.

'The greatest threat to our planet is the belief that someone else will save it.'

ROBERT SWAN[1]

Through the advancements in digitalizing supply chains, we are seeing the transformation of how companies operate and the impact this has on sustainability and social responsibility. In this chapter, we will explore the environmental and social implications that supply chain technology has, what best practices are being seen and the importance of these practices as companies pursue their agendas. It should be noted that globally, tough economic times, short-termism of investors and changing agendas of world leaders make these issues only harder to address.

REAL-WORLD EXAMPLE Procter & Gamble

Procter & Gamble (P&G) is the largest consumer goods company, founded in 1837, when a British candlemaker, William Procter, merged his business with James Gamble, an Irish soap maker in Cincinnati, USA. In 2025, they operated in 10 categories: fabric care, home care, baby care, feminine care, family care, hair care, skin and personal care, oral care, personal health care and grooming, with 65 brands sold in 180 countries and over 120,000 employees.[2] The company has been admired for many years for its supply chain initiatives and is recognized as one of four supply chain 'Masters' by Gartner,[3] with 132 manufacturing plants and 200 distribution centres.

Being a major force in the industry, the company faces ongoing scrutiny from regulators and consumers to lead the way in adopting practices that are sustainable and socially responsible. Over the years, it has committed to and achieved many sustainability goals, some of which are relevant to the use of supply chain technology.

It is upfront with providing information about its supplier list for commodities, specifically palm oil. This information is released biannually and covers 98 per cent of its global purchases. It also provides a list of its 1,352 mills within its supply chain, providing transparency to any stakeholder interested.[4] To support this transparency, it has used blockchain technology to track products from source to retailer so it can certify that its products are sustainably sourced. It supports its supply base with setting standards in relation to environment and ethics, which all parties within the ecosystem must abide by and share information about on their website. This helps it build trust with end consumers and ensures that all parties operating in the supply chain are paid for the activities they execute on P&G's behalf.

Another example is in relation to energy consumption because, as you can imagine, it consumes vast amounts of energy across its global assets, which contributes to its carbon footprint. To tackle this, it has invested significantly in AI and data analytics to firstly monitor energy consumption across its plants and deploy optimization techniques via AI systems to better understand its energy needs. Then armed with data, it examines how to reduce energy requirements by taking each part of the manufacturing processes apart to see if improvements can be made to reduce energy on production lines, for example and introduce automation where appropriate to improve energy saving.[5] Through these and many other initiatives using supply chain technology, it has been able to cut emissions across its operations by 60 per cent and still has ambitions to be net zero on emissions by 2040.[6]

Key agendas for the supply chain executive

When we think of the crucial elements that aid decisions for supply chain executives, we have already gone into detail on the pentagon of trade-offs in the early chapters of this book. Sustainability in the context of supply chain technology is the economic and environmental considerations that must be debated and agreed in the design, implementation and evolution of supply chain systems and processes. The overall focus of sustainable supply chain technology must be to minimize the impact supply chain operations have on the environment, economy and society, while still maximizing efficiency, effectiveness and profitability.[7]

From an emissions perspective, you will hear terms such as scope 1, 2 and 3, which we defined in Chapter 1 and have recapped here:[8]

- **Scope 1** – emissions that originate from a company directly such as manufacturing plants.
- **Scope 2** – emissions that originate indirectly through the purchase and use of electricity, fuel, steam, heating and cooling.
- **Scope 3** – emissions that a company is responsible for across its end-to-end supply chain from suppliers through to customers.

The hardest of these to control are naturally scope 3 because the other two you can manage and influence directly.

Intentionally, 'social' implications have been separated here, but in most instances, would be considered under the umbrella of sustainability. The reason for their segregation in the context of supply chain technology is to bring attention to the ethical and moral obligations that companies must adhere to so that their operations and practices promote and protect the human rights of all individuals and communities involved in their end-to-end supply chains. In practice, this should include visible and documented agreements in relation to fair labour practices, safe working conditions and respect of human rights throughout the supply chain.

Having digital evidence of how a company operates within a larger ecosystem can help create new opportunities to improve cost positions, bring assurance and/or pinpoint issues and show where weaknesses exist.

When comparing trends across a variety of supply chain research,[9] there is no getting away from the importance of sustainability and social responsibility when companies decide what to do or what not to do. In February 2025, the Trump administration began to backtrack on climate action in the USA,[10] meaning that net zero targets that were behind already[11] moved further out of reach. In 2025, BP reduced its investments into clean energy, reversing its position back[12] to a pre-Bernard Looney strategy.[13] Bernard Looney, the former CEO of BP, had a strategy that put the importance of sustainability at the heart of a fossil fuel company.

However, there is a strong need for companies to understand their sustainability and social responsibilities as regulation and compliance is ever-increasing, especially in the EU,[14] although implementation dates have continued to slip.[15] Companies cannot turn a blind eye to failing standards because the consequences are severe and damaging. For example, research by Fitch Ratings suggests the number of goods produced using child labour in 2020 in agriculture, separately, is still approximately 20 per cent. It also stated that while legislation exists in many western countries, it is good for raising awareness, but it lacks legal force and penalties for companies to be proactive with them.[16]

REAL-WORLD EXAMPLE Nike

In 2001, it emerged that Nike's suppliers were using sweatshops in developing countries.[17] The backlash from the public led to its reputation falling among even 'die hard' fans of the brand, with customers and stakeholders vocalizing how disappointed they were and questioning its ethics. In the face of criticism, Nike took proactive steps to enhance its supply chain responsibility by:

- establishing rigorous labour standards to provide safe working conditions and fair treatment for workers in its end-to-end supply chain
- providing transparency to stakeholders about its operations and practices
- developing a code of conduct for its suppliers, outlining the expectations for responsible business practices and labour standards
- improving working conditions throughout the supply chain, ensuring a safer and healthier environment for workers

- embracing corporate social responsibility by prioritizing initiatives to rebuild its brand and restore consumer trust, demonstrating a commitment to responsible business practices.

By actioning these steps, Nike was able to course correct, gaining better control and understanding of what was happening in its supply chain to regain brand loyalty.[18]

A best practice approach was demonstrated by Paul Polman, the CEO of Unilever for 10 years, who was transformational in his approach and vision for a sustainable and responsible business, with a focus on reducing environmental impact and improving social outcomes.[19] His initiatives included:

- **Unilever Sustainable Living Plan (USLP)**, a 10-year plan to reduce Unilever's environmental impact and improve social outcomes.
- **Sustainable agriculture**, where Unilever committed to sourcing 100 per cent of its agricultural raw materials sustainably by 2020.
- **Renewable energy**, where the company aimed to power 100 per cent of its operations with renewable energy by 2030.
- **Waste reduction**, which aimed to halve its waste footprint by 2020.

The point of these examples is to set the tone that companies are waking up to the realization that they cannot just be interested in their part of the supply chain, predominantly more so in Europe than other regions at this time due to the increasing regulation.

In most instances, they are the largest player in the end-to-end supply chain, so they have the most to lose if a 'bad' actor operates within the ecosystem. This has led to having to answer more difficult questions in relation to supply chain. A study by IBM in 2023 stated that only 38 per cent of businesses were tracking scope,[20] but technology is now slowly helping to aid companies as they try to understand their impact (although many companies are still collating this information on spreadsheets).[21] Once companies are comfortable with an understanding of their impact, the conversation should quickly pivot to what are they going to do about it to minimize the impact going forward.

Sustainability

Diving into sustainability specifically in relation to supply chain technology, there are four key areas that focus needs to be given to: environmental, economic, technological and social. Here we will consider how, by using technology, we can report on these areas, but more importantly, action ways to improve these for the benefit of the company and to protect our futures.

From an environmental perspective, a company should understand and minimize its supply chain's footprint of its operations, considering greenhouse gas emissions, energy consumption, water usage and waste.

Economic sustainability involves protecting the viability and profitability of operations while minimizing costs and maximizing value for stakeholders in alliance with environmental and social initiatives. For example, in routing, this involves looking not just at the obvious time, distance and cost but also at the type of vehicle, the type of fuel and understanding the impact on the driver to come up with a more informed route that balances these factors.

From a technology angle, companies must consider how to develop and implement supply chain technology that provides them with adaptability, scalability and resilience while minimizing energy consumption.

Lastly, from a social perspective, the company should promote labour practices that are fair and equitable, making sure workers are operating in safe conditions and are treated fairly in relation to their rights, dignity and well-being. Human rights should be at the heart of a company's values and ignorance is not an excuse that customers accept. As supply chains stretch across the globe, the use of forced labour, child labour and human trafficking must be eradicated. Furthermore, correct health and safety measures should be in place to protect workers from operating in unsafe environments and they should have the necessary resources and training. Companies and their supply chains should be contributing to the economic and social development of the communities in which they operate, creating jobs, education opportunities and engagement with others. This all should lead to companies providing transparency and being accountable for not just their piece of the supply chain but through orchestration with all parties within it to provide this level of care for others with regular audits, assessments and reporting where the strength of technology must be deployed.

Sustainable supply chains should provide efficiency gains by reducing energy consumption, helping reduce waste and costs.[22] They should provide companies with the ability to demonstrate transparency and help pinpoint risks with increased visibility of the supply chain. Resilience will be improved as companies are able to see and respond to disruptions much sooner and mitigate risks in near real time, leading to improved accountability and trust.

With these elements in place, better decision-making should be the outcome with more informed decision-making capability.

From a social perspective, using technology will help companies with their reputation and brand value because they will have more control over their ecosystem from source to customer. With that comes trust because if they are socially responsible and sharing their data with relevant parties' customers, supply base, investors and employees, they in turn will get behind the company mission. Risk will be proactively managed whether from a reputational perspective, reducing financial losses or aligned to regulatory compliance. Finally, it will help decision-making having this information to hand, which only technology can provide when operating with a complex supply chain.

How supply chain technology helps these agendas

We can quickly get fixated on technology use in supply chains as being for a specific reason or goal and miss the impact it can have beyond. Environmental sustainability and social responsibility could be deemed a secondary concern or a benefit of these advancements in the use of technology in the supply chain. However, thanks to the influencers in this space, including Al Gore (the US politician and environmentalist), Paul Polman (ex-CEO of Unilever) and Yvon Chouniard (founder of Patagonia), there has been an emergence of many new companies whose sole focus is on looking at supply chains through a sustainable lens. Thought leaders such as Zero100, whose mission includes shining a light on the impact companies' supply chains are having in this space, as well as in the wider technology advancement space of supply chain,[23] have also been impactful.

Solutions that help companies with sustainability are focused primarily on capturing data from environmental, social and governance perspectives and creating a single system of records. This helps with reporting internally and to external bodies but also means that any actioning is done from this, rather than by using technology to help analyse further and pinpoint areas of concern (e.g. high emission factories) or where gaps exist to be actioned on. The growing ability of analytics, as we have already explored in detail in Chapter 5, has enabled more capabilities to exist, which in time will be very helpful in this area.

REAL-WORLD EXAMPLE BAM Clothing

BAM Clothing, founded in 2006,[24] is an online clothing company that sells leisure and fitness clothes using bamboo (one of the world's most sustainable crops)[25] as a fabric. From the outset, its focus has been on sustainability and transparency through the 'crop to wearing' supply chain.

In 2021, it began an initiative to map its entire supply chain back to its sources, which included tracking down and understanding its suppliers' suppliers. Then it engaged GreenStory,[26] a software company that evaluates product footprints, which conducted life cycle analysis (i.e. measures the environmental impact of a product through every phase of its life)[27] using its technology to review emissions, energy efficiency, blue water consumption and understand the environmental footprint of 1kg of input (crop / initial fibre) to reaching the customer. Once BAM Clothing knew its position, it then looked at ways to reduce its environmental impact further.

It also spent considerable time with its suppliers to ensure they were adhering to ethical and environmental standards. Within this process was the need for transparency, building trust and accountability among partners.

Although this shows the effort required to determine your environmental impact, it also shows how using technology can assist you in achieving this understanding to abide by regulations and build trust with your customers.

Environmental sustainability

The increased usage of technology in supply chains has led the rise in energy consumption. This has a direct impact on greenhouse gas emissions, which are only increasing more and more with the need for more data centres. Very few of these are energy-neutral like Virtus;[28] the majority, including Amazon, Google and Apple, are nowhere near achieving that.[29]

DHL plans to reduce its scope 1 and 2 by 42 per cent by 2030 and be net zero by 2050[30] by reducing fuel consumption, electrifying transport and overall lowering its emissions. Across the world, investments in solar, wind and other renewable energy sources are being made to power supply chain operations. In manufacturing plants, many novel ideas are being deployed, which are focused on energy efficiencies from basics such as energy-efficient lighting, to 'smart' heating, ventilation and air-conditioning systems, which are managing temperatures and providing better conditions for employees, leading to reduced energy consumption and lower operating costs. For example, Coca-Cola in its Austrian manufacturing plant, has reduced its emissions by 50 per cent since 2010, deploying such initiatives.[31]

Passing laws is one approach to get companies to change but this can take considerable time. The challenge that companies face is how to marry sustainability with being cost-neutral (or better). One approach that is slowly gaining traction through the supply chain is the use of 3D printing, which means specific components can be made close to where they are combined for manufacturing into finished goods (even within the same facility), therefore eliminating the logistics element. For example, it can reduce the carbon footprint from 14 tons to 0.8 tons compared to traditional supply chains.[32]

As covered in depth in Chapter 4, visibility and transparency play a crucial part within sustainability initiatives and require the use of technology. This is both to track their impact at different stages of the supply chain and for the ongoing monitoring providing surface data to then act on. Examples of this would be the monitoring of a company's carbon footprint, understanding the impact the company and its supply base have and then innovating new ways to reduce its footprint without impacting the overall service to customers. In many parts of the world, technology and data is also forming the basis of statutory reporting, which helps stakeholders such as governments and investors decide how well a company is managing this as part of its overall performance.[33]

Social responsibility

Social initiatives in the supply chain can also benefit from the use of technology. For labour practices, technology can be used in manufacturing plants and warehouses to track working hours and wages, as well as records of employment that capture workers' data. This can then monitor human rights and help reduce the risk of abuse in supply chains that are connected end-to-end. Regular audits and certifications backed by technology can provide protection and compliance to labour standards required by customers.

For communities, technology assists with their engagement and development. Companies can invest in local economies and provide jobs to develop sustainability initiatives that technology can monitor. Also, technology can be used to train and build skills in the workforce and provide workers with consistent standards aligned with the overall company standards.

We have discussed some of the sustainable supply chain technologies that are applicable for environmental sustainability and social responsibility already, such as the use of blockchain to promote transparency and accountability across global supply chains. IoT devices can be used to provide near real time data on company working conditions, ensuring staff are treated aligned to labour laws and human rights.[34] IoT devices can also provide data on energy consumption and waste. Over time, all this data can be tracked, then using technology, supported by human input and AI input, glean insights and improve decision-making.

Ideas to consider

Every company faces sustainability and social challenges in relation to its supply chain, so there is a significant need to innovate ways to overcome them, to which technology can be a great help. The key point is to keep a continuous improvement mindset within the different parts of your supply chain and encourage one from your external partners to find ways to improve the supply chain and its performance. Here are some suggestions that I have gained from assisting clients in this area:

- Have in place framework agreements, detailing suppliers' specific ways of how they are prioritizing sustainability and social initiatives, as well as, at an overarching level, ways of how they conduct business with your company. This should include the suppliers being compliant with internationally set standards, such as ISO in relation to environment management, health and safety and energy management, for example. It is only by companies setting standards for themselves and others who do business with them that the overall ecosystem improves and it should not be one way: engagement and open-mindedness to help each other is crucial.
- Conduct regular audits of other parties, not just as 'tick the box' exercises but to also share best practices and be part of any solution development.
- Invest in technology that helps to reduce energy consumption, emissions and waste. Looking at the bigger picture on these topics, being willing to consider investments that may take over usual payback periods but address these issues is important. Opportunities to understand what others are doing should be embraced. For example, attend suitable conferences and engage (where acceptable) with other customers of suppliers because knowledge sharing will benefit all.

- Encourage transparency and accountability across the end-to-end supply chain using some of the technologies discussed to monitor and report how parties are doing and to provide guidance on expectations. There are many different solutions in the market that can help companies get an improved understanding of their position and this is the first step on the journey towards sustainable supply chains. The challenge is then having the capability to provide scenarios of environmental impacts so that more informed decisions can be made. Take, for example the 'last mile' – MIT found that if you ask the consumer: 'Do you need the product the next day or are you willing to wait?' by explaining the impact of that choice through an environmental lens, the number of consumers who were willing to wait increased.[35] All of the analyses done to support this are using supply chain technology and it can capture those types of examples and provide a better understanding of carbon emissions and how they can be reduced.
- Engage with stakeholders in the community to promote sustainability and social responsibility. This is important for companies to prosper in different geographies. Using supply chain capabilities in the community can help invigorate them, such as by providing tools to help farmers have more control over crop cycles, understand the quality of their products when brought to co-operatives (e.g. cocoa, coffee) and then use mobile applications to buy and sell. Also, using a mixture of less energy consumption and renewables in company assets can lead to better use within communities or provide more energy for their use. For example, Starbucks and its ethical sourcing promise aligned to Coffee and Farmer Equity (CAFÉ) practices.[36]
- Have a comprehensive supply chain strategy that prioritizes sustainability and social responsibility. Thinking through the pentagon of trade-offs will improve decision-making.

EXPERT INSIGHTS

When developing a platform aimed at promoting sustainable products and educating companies about their impact, you first need to reflect on their values (Why are they pursuing such an agenda?) and help them to understand that the technology and data inputted into it need to be more than for compliance's sake. It needs to be a topic that informs them of how they are performing but more importantly, what they need to focus on to drive sustainable practices into their supply chain. Those companies that have done this achieve better outcomes.

The importance of data quality and having confidence in it is a must – if you don't, you can quickly waste significant capital on initiatives with negligible impacts. Conducting a life cycle analysis and then using technology to capture relevant data is one approach and while it can be time-consuming to engage many parties in the supply chain, it is a practical first step.

Consumers can be sceptical of brands that talk about sustainability, questioning whether they are doing it to attract customers or because of regulatory requirements (i.e. stick-and-carrot). There are a variety of perspectives, but typically, the small to medium-sized companies like Patagonia, BAM and Allbirds are driven by leadership to have sustainability at the core of their being and don't necessarily do it for financial benefit. In some cases, sustainability is their reason for existing (similar stories exist with manufacturers as well). Implementing sustainability initiatives becomes a challenge in relation to global scalability within organizations that are trained with business cases and financial metrics that are used to decide whether to invest. Also institutes like the World Bank and governments (in early 2024) altered their understanding of the carbon footprint, what activities are driving it and the data behind it to inform decisions they want to make.

Strong regulations are necessary for companies to maintain their sustainability efforts in the long term. The role of technologies such as digital product passports (DPPs) in identifying supply chain issues and driving accountability among suppliers is upon us (as discussed in Chapter 4). What we need to consider is that consumer attention spans are short, so unless companies keep sustainability in focus through their messaging, it might become more of a fad than part of how they conduct their business. To overcome this, again the company will need to have good data with companies willing to spend the time to capture it and have their upstream parties bought into providing the relevant data and keeping it refreshed. To overcome the tension between parties, it is essential that they also collaborate with governments to create a sustainable future following consistent standards.

Existing technology for sustainability has been criticized for lacking accurate data collection and validation processes. The role that AI can play in this today is mostly focused on processing and validating data. For example, AI could determine whether a manufacturing site that has been audited is a fair comparison to one that exists within a specific supply chain. Doing this accelerates the path to using data and technology in tandem as an organization's baseline on which it can then act. Many companies are under the illusion that their sustainability data is more mature than it is, leading to misguided investments. Further initiatives that AI can support are in relation to training, which can help parties in the supply chain as they go about capturing relevant information. They can be guided in relation to specific tasks rather than generic training delivered via videos, for example.

While technology has the potential to drive improvements in sustainability, it is fair to say that sustainability is not the major focus within the topic. With analytics (as discussed in Chapter 5), we already have a blueprint of what to do with sustainable data once we have confidence in it. Technologies that already exist in this space, such as blockchain, can help with transparency, but the system in its current form does not allow a detailed review of data and parties involved from an external third party to ensure all the data is valid. It is more akin to a ledger to protect and reward parties that engage and execute against the ask.

So, while there are varying levels of confidence, it is encouraging to see that there is eagerness to build future sustainable supply chains with the increasing abilities provided by technology to meet that need when data catches up.

Conclusion

The two topics of environmental sustainability and social responsibility don't just affect us in the workplace today but are actions that will have significant ramifications on many future generations. If companies can weigh up decisions using the supply chain pentagon (Figure 3.2) that considers service, cash, cost of capital, resilience and sustainability, and reflect on the impact they are having, there will continue to be many more innovations in the supply chain, as technology deployment will help reduce emissions and waste, minimize costs and improve the communities in which the supply chains operate.

The examples used in this chapter show how, in a relatively short period of time, companies can understand their impact on the environment and society and can then turn the insights into initiatives to improve (or at best eliminate) activities that have negative impacts. By focusing on energy, emissions, waste and treating labour fairly and equitably, the supply chains of the future will be robust and more transparent, which should bring the standards for civilization up as well. Continuing to innovate will make the technology that enables these ideas cheaper and with that, become more appealing and reduce barriers for organizations and their partners to implement across their global supply chains.

Endnotes

1 Swan, R and Reavill, G (2009) *Antarctica 2041: My Quest to Save the Earth's Last Wilderness*, Broadway Books
2 P&G, P&G History, P&G, www.pg.co.uk/pg-history (archived at https://perma.cc/YH2M-3V62)
3 Lavelle, J and Ruane, B (2024) Gartner announces 20th annual rankings of the Global Supply Chain Top 25, Gartner, www.gartner.com/en/newsroom/press-releases/2024-05-22-gartner-announces-20th-annual-rankings-of-the-global-supply-chain-top-25 (archived at https://perma.cc/BBQ9-YXEX)
4 P&G (2025) P&G Palm Oils Mill List, P&G, https://www.pgchemicals.com/downloads (archived at https://perma.cc/69CF-6AQY)
5 MAD / Insight (2024) How AI is helping P&G ensure its packaging is reusable or recyclable, MAD / Insight, www.madfestlondon.com/insights/articles/dragonfly-article (archived at https://perma.cc/EB86-AA9J)
6 P&G (2024) Environmental – Climate, P&G, www.pginvestor.com/esg/environmental/climate/default.aspx (archived at https://perma.cc/2XBW-KS72)
7 Hariyani, D, Hariyani, P, Mishra, S and Sharma, K M (2024) A literature review on green supply chain management for sustainable sourcing and distribution, *Waste Management Bulletin*, 2 (4) www.sciencedirect.com/science/article/pii/S2949750724000993#:~:text=Closed%2Dloop%20supply%20chain%20(CLSC,et%20al.%2C%202019) (archived at https://perma.cc/UMA2-DPYH)
8 Carbon Neutral (2024) Scope 1, 2, and 3 emissions explained, Carbon Neutral, www.carbonneutral.com/news/scope-1-2-3-emissions-explained#:~:text=These%20are%20the%20emissions%20that%20come%20from%20the%20waste%20disposal,%2C%202%2C%20and%203%20emissions (archived at https://perma.cc/8EJQ-P7JF)
9 Kinaxis (2023) Connecting supply chains to corporate strategy, Kinaxis, https://kinaxis.com/en/blog/connecting-supply-chains-corporate-strategy?language=de#:~:text=To%20do%20this%2C%20organizations%20need%20to%20reconsider,cost%2C%20cash%2C%20service%2C%20environmental%20impact%2C%20and%20resilience (archived at https://perma.cc/2E7J-R6YM)

10 Milman, O (2025) Scientists brace 'for the worst' as Trump purges climate mentions from websites, *The Guardian*, www.theguardian.com/us-news/2025/feb/04/trump-climate-change-federal-websites (archived at https://perma.cc/2ATF-VANZ)

11 University of Oxford (2024) Redefining net zero will not stop global warming, new study shows, University of Oxford, www.ox.ac.uk/news/2024-11-18-redefining-net-zero-will-not-stop-global-warming-new-study-shows#:~:text=within%20a%20generation.-,Professor%20Myles%20Allen%2C%20Department%20of%20Physics.,al%2C%20and%20Gregory%20et%20al (archived at https://perma.cc/Q2K2-BHK8)

12 Reed, S (2025) BP to focus more on oil and gas in strategy 'reset', *The New York Times*, www.nytimes.com/2025/02/26/business/bp-oil-gas-renewables-strategy.html (archived at https://perma.cc/AV4S-AXFD)

13 BP (2020) BP sets ambition for net zero by 2050, fundamentally changing organisation to deliver, BP, www.bp.com/en/global/corporate/news-and-insights/press-releases/bernard-looney-announces-new-ambition-for-bp.html (archived at https://perma.cc/9PLZ-3QY5)

14 European Commission (2025) Corporate sustainability and responsibility, European Commission, https://single-market-economy.ec.europa.eu/industry/sustainability/corporate-sustainability-and-responsibility_en (archived at https://perma.cc/LZ3X-ZV36)

15 The Biodiversity Consultancy (2025) What businesses need to know about CSRD and CSDDD omnibus changes, The Biodiversity Consultancy, www.thebiodiversityconsultancy.com/insights/what-businesses-need-to-know-about-csrd-and-csddd-omnibus-changes/#:~:text=The%20proposed%20changes%20in%20the,Grant%20Rudgley%20(Nature%20Finance) (archived at https://perma.cc/5UYF-3WBY)

16 Chike-Obi, N and Tang, M (2021) Modern slavery and labour risk in global supply chains, Sustainable Fitch, Fitch Ratings, www.sustainablefitch.com/_assets/special-reports/modern-slavery-labour-risk-in-global-supply-chains.pdf (archived at https://perma.cc/5MAW-BV7K)

17 Wazir, B (2001) Nike accused of tolerating sweatshops, *The Guardian*, www.theguardian.com/world/2001/may/20/burhanwazir.theobserver (archived at https://perma.cc/VV89-THAJ)

18 Carlson, D (2025) Nike, Inc., Britannica Money, www.britannica.com/money/Nike-Inc (archived at https://perma.cc/2YU9-ECGH)

19 Oxford (2024) Case study: Unilever's 'Sustainable living plan', Oxford Executive Institute, https://oxfordexecutive.co.uk/case-study-unilevers-sustainable-living-plan (archived at https://perma.cc/Q9ZC-S6LC)

20 Unglesbee, B (2023) Only 38% of businesses are tracking Scope 3 emissions: IBM report, Supply Chain Dive, www.supplychaindive.com/news/Scope-3-carbon-emissions-tracking-supply-chains-IBM-survey/694956 (archived at https://perma.cc/9YCD-MHDT)

21 Cajic, M (2024) The dangers of using a spreadsheet for carbon accounting, Avarni, www.avarni.co/news/the-dangers-of-using-a-spreadsheet-for-carbon-accounting (archived at https://perma.cc/69VN-XHK3)

22 Sánchez-Garcia, E, Martinez-Falcó, J, Marco-Lajara, B and Manresa-Marhuenda, E (2024) Revolutionizing the circular economy through new technologies: A new era of sustainable progress, *Environmental Technology & Innovation*, 33, www.sciencedirect.com/science/article/pii/S2352186423005059 (archived at https://perma.cc/RW56-M964)

23 Zero100 (2025) Mission, https://zero100.com (archived at https://perma.cc/GQT6-QRMF)

24 BAM (2025) About us, BAM Bamboo Clothing, https://bambooclothing.co.uk/bam-story (archived at https://perma.cc/DS3K-JP4G)

25 GreenStory (2022) In conversation with Merryn Chilcott, BAM Bamboo Clothing, GreenStory, www.greenstory.io/blogs/in-conversation-with-merryn-chilcott-bam-bamboo-clothing (archived at https://perma.cc/AH4S-X8WB)

26 Product DNA (2025) BAM Bamboo Clothing, https://bambooclothing.co.uk/product-dna-landing (archived at https://perma.cc/A6X6-EJHA)

27 Quist, Z (2024) Life cycle assessment (LCA) – everything you need to know, Ecochain, https://ecochain.com/blog/life-cycle-assessment-lca-guide (archived at https://perma.cc/39SW-S7D4)

28 Virtus (2024) Green 100% renewable energy, Virtus, https://virtusdatacentres.com/why-virtus/green#:~:text=Green%20100%25%20Renewable%20Energy,our%20fuel%20mix%20100%25%20green (archived at https://perma.cc/4W64-G7J8)

29 O'Brien, I (2024) Data center emissions probably 662% higher than big tech claims. Can it keep up the ruse? *The Guardian*, www.theguardian.com/technology/2024/sep/15/data-center-gas-emissions-tech (archived at https://perma.cc/3TC6-XTK9)

30 DHL (2024) PPN 06/21 – Carbon Reduction Plan, DHL, www.dhl.com/content/dam/dhl/local/gb/dhl-supply-chain/documents/pdf/dhl-supply-chain-uk-ppn-carbon-reduction-plan-march-2024.pdf.coredownload.pdf (archived at https://perma.cc/M2TD-DCGJ)

31 Siemens (2020) Coca-Cola is boosting energy efficiency, Siemens, www.siemens.com/global/en/company/stories/industry/energymanagement-coca-cola.html (archived at https://perma.cc/4N4E-5Q6B)

32 Markforged (2024) 3D printing and the environmental impact of manufacturing, Markforged, https://markforged.com/resources/blog/3d-printing-and-the-environmental-impact-of-manufacturing#:~:text=3D%20printing%20reduces%20the%20environmental,manufacturing%20whatever%20components%20they%20can (archived at https://perma.cc/J9GJ-6UZB)

33 Anderson, I (2024) Three ways to harness technology and data to deliver your sustainability strategy and reporting, www.pwc.co.uk/services/esg/harness-technology-and-data-deliver-sustainability-strategy-and-reporting.html (archived at https://perma.cc/ZX9Y-XPNG)

34 Wall, D (2023) Why ESG is important for big players in the IoT world and how IoT can help to achieve ESG goals, *IoTNow*, www.iot-now.com/2023/01/27/127207-why-esg-is-important-for-big-players-in-the-iot-world-and-how-iot-can-help-to-achieve-esg-goals/#:~:text=Environmental%20monitoring:%20IoT%20sensors%20can,working%20conditions%20for%20their%20employees (archived at https://perma.cc/XN8J-V3LJ)

35 Martinez, J V, Rajagopalan, S, Fransoo, J, C and Valenzuela-Ocana, K (2024) The effect of framing the environmental impact of fast shipping on nudging consumers to opt for slow home deliveries: Evidence from the field, MIT Center for Transportation & Logistics Research Paper No. 2024/008, https://papers.ssrn.com/sol3/papers.cfm?abstract_id=4936719 (archived at https://perma.cc/2T48-456C)

36 Starbucks (2024) C.A.F.E. practices: Starbucks approach to ethically sourcing coffee, Starbucks, https://about.starbucks.com/press/2024/cafe-practices-starbucks-approach-to-ethically-sourcing-coffee (archived at https://perma.cc/8V4Q-M7Y7)

PART THREE
Implementing digital technologies in supply chain

Building a business case for supply chain technology investment

09

This chapter will discuss the importance of building a business case for supply chain technology investment and how to conduct an ROI analysis and cost-benefit analysis for this purpose.

> **CHAPTER AIMS**
>
> - Learn how to build a compelling business case for supply chain technology investment.
> - Understand the key components of a successful business case, including ROI analysis and cost-benefit analysis.

'We do not have a fear of the unknown. What we fear is giving up the known.'

ANTHONY DE MELLO[1]

In this part of the book, we will first look at the importance of a business case to support supply chain technology investments. From there, we will look at how to measure success from these investments before looking at the role of people and the importance of adoption. We will then look at how to succeed at implementation before concluding Part 3 by looking to the future.

EXPERT INSIGHTS

Some sage advice from other experts is sprinkled through this chapter and this is where we will start. At the outset of any project, we have already touched on the necessity of a business case and in this chapter, further insights are provided.

The first point is that there is no one right approach to the mechanics of putting the business case together. What there is agreement on is having a common view of the end state and what we expect the technology to enable us to do that we cannot today or to modernize legacy systems. Misalignment on expectations at this stage usually doesn't prevent progress but can often lead to issues later in the life cycle of the project. So, team members sharing a clear understanding of the business rationale and transformation is a must early in the process.

From a benefits perspective, there is agreement that simple trumps all, also being very careful that benefits identified do not cannibalize other benefits identified. For example, if you are claiming inventory benefit, make sure that you can follow it to a measurable point because reducing inventory reduces working capital and costs related to the movement of the product. Still, it could lead to higher indirect costs, such as cost of sale. This is because not buying aligned to contract volumes means a higher cost to serve for the rest of the portfolio (since you might have fixed costs on lower capacity) or more expediated shipments (i.e. road to air) and higher lost sales because you don't have inventory when a customer wants it.

Also, there are two types of technology investments: strategic (driven by long-term plans) and reactive (responding to immediate operational needs like a new warehouse). Both require different approaches but start with a gut feeling and benchmarking potential solutions. Supply chain professionals often rely heavily on solution providers to help build business cases, as they may lack the expertise to do so internally.

There are many reasons why companies need to implement supply chain technology solutions, from vendors sunsetting their solutions, to new capabilities coming to market that include AI, meaning the technology estate is constantly modernising. The ongoing vast improvement of computational power is a major driver whereby we can run much faster sophisticated optimization algorithms with more constraints, for example.[2] Also, data (internal and external) that can influence demand or supply is now more accessible via data warehouses and data lakes. The user interface (UI) has gone through a revolution on how to visualize data and create alerts for interactive dashboards. Developments in AI are opening opportunities for automation, collaboration and decision support, which we will look at in more detail in Chapter 13. Whatever your specific reason for implementing a supply chain technology solution, there are key steps you must take to try to make sure your project doesn't fail, since sadly there are graveyards full of technology projects that were started with good intentions but fell apart, went over budget and missed timelines. There are even stories of companies that took so long to implement a solution that by the time it was done it was out of support, so they had to move on to the implementation of the next version.

Statistics (and there are many in relation to IT failures) provide grim reading: 70 per cent of efforts fail[3] and McKinsey says that average IT projects run 45 per cent over budget, 7 per cent over time and deliver 56 per cent less value than expected.[4] Therefore, the challenge for companies is finding how they can be in the minority of these figures and make their supply chain management IT transformations a success.

Of course, companies don't want the outside world knowing about the failures. However, if a company has the misfortune to be involved in one that is customer-facing, it can quite quickly become front page headlines – just see The White Company,[5] End Clothing[6] and PepsiCo[7] who have all had to explain publicly what went wrong with technology transformations in their supply chains.

One company that showed the knock-on effect of an unsuccessful supply chain technology investment was Levi Strauss, which, during an ERP system upgrade, ran into significant problems when one of its customers needed the system to interface with their supply chain management system. At system switchover, they were unable to fulfil orders, having to take three US distribution centres offline for a week. This meant they incurred a charge of $192.5 million against earnings. From an initial budget of $5 million to implement the technology, the impact was almost a $200 million loss. This event, coined as 'black swan' (rare, high impact), seemed in hindsight, rather predictable.[8]

Another example is Nestlé. Its health science division, which had grown through the acquisition of several companies, all had their own systems, which were deemed outdated. It consolidated these systems into a modern automated system but had not appreciated the complexity of consolidating many different systems and SKUs. This led to not having stock available with a short-term impact on business performance: a reduction in operating margin of 1.6 per cent for the quarter.[9]

Investments in supply chain management technology can have a significant impact on a company's bottom line and commitment of resources. Therefore, the need for a robust business case is required to justify the decision and should be referred to and updated throughout the life of a project.

The need for creating a business case

Investments in supply chain management technology are typically material in cost and come with risk, require resources and time across different functions of the business and can involve gaps in time between the investment and seeing the benefits. They can also have several phases to go through and can have a multi-year timeline depending on the technology being invested in. Many of the following points are applicable beyond supply chain technology investments, however I want to take the opportunity to go back to basics and build the case 'bottom up'.

A business case in essence is a structured approach to evaluating the potential benefits and costs of a supply chain technology investment (in our case).[10] It helps companies to identify the key drivers and objectives of an investment, quantify the potential benefits and costs, evaluate the ROI and payback time and develop an implementation plan. In many instances, an initial business case looking for investment to conduct a pilot (trial) can be formed, which when completed, then builds further support (and evidence) to proceed to a full business case.

The business case needs to:

- address many different stakeholders' requirements
- be compelling
- be easily digestible
- state the benefits of the new technology solution
- acknowledge the assumptions where facts cannot be determined or there are dependencies on certain other business outcomes which may not occur (e.g. business growth)
- lay out the expected return on investment (ROI) that the new technology solution will provide.

The business case is critical because it helps decision makers prioritize what is important to the company. It achieves this for a variety of reasons, firstly helping to justify the investment through concise and compelling arguments for why the technology is needed. To secure funding, the art of storytelling (where visual aids should be employed to simplify complex information) and providing a well-thought-out business case that demonstrates indicative ROI is usually required. It can help corral stakeholders around a common goal and helps focus them on working towards the same objective. Using a business case, it provides decision makers with an ability to evaluate different options (considering multiple scenarios and sensitivity analysis) and choose the appropriate solution for their company. Usually, a senior leader within the supply chain function writes the business case, with the CSCO as the main sponsor.

EXPERT INSIGHTS

As one supply chain director I spoke to puts it, any business case for supply chain technology investment should stem from a clear and concise business need rather than trends or aesthetics. The investment must be tied to either opportunities or problems within the supply chain context and be considered with the maturity of the current operation and future goals. Technology should be viewed as an enabler within a comprehensive supply chain operating model (including people, process and data) rather than a standalone solution.

The key components of a business case should be an understanding of the current state of performance, the operational ecosystem and identifying future opportunities.

Technology investments often involve behavioural changes and require a holistic view of the ecosystem, making it challenging to pinpoint discrete benefits. Benefits should be viewed collectively rather than in isolation, as the success of technology relies on the evolution of organizational processes and capabilities.

The business case must be practical: don't over-complicate the process, compelling narratives and a clear understanding of the transformation required are more critical than formal structures. Many technology projects fail to meet their business cases, often due to a lack of focus on critical enablers and the complexity of stakeholder engagement (which we will get into in Chapters 11 and 12).

What should be in a successful supply chain transformation business case?

In this section, we will work through the key elements that should be documented in a business case (see Figure 9.1), with supply chain examples. Following this section, I will work through a more detailed example that reflects some business cases I have worked on.

- **Business problem statement** – this explains in concise, simple language the context and what the business problem or opportunity is that the technology investment aims to solve. It highlights any pain points and challenges that the investment should solve as well.
- **Solution overview** – this summarizes the proposed technology solution and should include outlining the key features and functionalities of the solution, explaining how, by implementing this technology, it will address the business problem or opportunity.
- **Benefits analysis** – the analysis should explain the potential benefits of the technology investment, quantifying it using a variety of metrics which should include ROI, payback period and net present value (NPV). Levers that should be assessed to identify opportunities include revenue growth, improved customer experience, cost savings and efficiency gains (and help transition towards sustainability). The key point here is to highlight the metrics that matter to the stakeholders in clear headings; these may vary between companies but engaging with finance colleagues will help to guide the business case owner to the appropriate benefits to communicate. This section should not get bogged down in having to justify and show calculations; these should be laid out in later sections or appendices.

Figure 9.1 Business case overview

Business case pack

- Business problem statement
- Solution overview
- Benefits analysis
- Cost analysis
- Financial analysis
- Implementation plan

- **Cost analysis** – the analysis should detail the total cost of ownership of implementing and operating the technology solution, plus lay out any potential cost savings or reductions. Examples of costs that should be included are hardware and software costs, implementation and integration costs, licence costs, ongoing maintenance and supporting costs as well as outlining cost savings or reductions that are linked to these costs. Like benefits, these should be headline-driven with supporting materials where appropriate. In most instances, it makes sense to split one-off expenditure and recurring expenditure. One cost element that is often neglected is related to change (which we will explore further in Chapters 11 and 12). During project implementations, the workload will temporarily increase while running the current business, where additional effort will be needed to develop and test the new solution and support rollouts. In addition, the new technology will bring new skill requirements with the consequences that some of the current team within the function may become redundant and new roles may need to be created and filled.
- **Financial analysis** – this shows the calculations and assumptions linked to the ROI of the technology investment, the NPV and detailing the payback period. This should be done in collaboration with finance colleagues. The following formulas are the standard approach to do doing this.
  - **ROI** = (gain from investment – cost of investment) / cost of investment then multiplied by 100 to convert into percentage
  - **NPV** = Projected cash flow for each time period (typically annual) / (1+ discount / hurdle rate desired by company) calculated for each year the project is expected to incur costs and benefits
  - **Payback period** = cost of investment / annual gain from investment

Knowing the NPV, alongside ROI, is helpful since ROI doesn't consider the time value of money – the future value may be of less value than the present value. ROI should be positive; if it is not, why are you proposing the solution? This is where I have seen technology business case proposals for ERPs on many occasions come unstuck. You start seeing the business cases populated with supply chain benefits, when in many instances, the benefits are driven primarily from a finance and data perspective. ERP modernization focuses on the record of transaction so the benefits they provide the supply chain may not be so considerable from that perspective alone. Arguably, these are foundational systems of the technology stack that enable supply chain benefits from technology capabilities deployed on top of them rather than via them. The exception to this would be unless you are moving from a legacy (i.e. old) system to a modern one.

Using NPV, you are assuming that the technology investment will provide returns of a certain amount in a future time and you are trying to ascertain what that value is worth today. Most companies have 'hurdle' rates in place, so if investments don't meet the agreed criteria, they are rejected or further analysis is required to justify it. Furthermore, on payback, technology solutions, like other assets, have a shelf life, so suggesting 20 years payback will not be appealing to decision makers (who are typically C-suite) unless in relation to warehouse ASRS, for example, which will be operational for a considerable time. Getting an idea of appetite from the finance department should assist and guide the owner of the case. I worked with one client who demanded payback of two years or less, which encouraged the supply chain team to design and implement solutions to achieve this threshold. However, once they had done this for a few years, they realized they would gain more ROI by doing 'surgery' rather than putting 'plaster upon plaster' in their warehouse operations. To achieve this, they had to change their approach to influence and gain support for a longer payback period. What helped them in this instance was the confidence they had provided stakeholders that they could deliver projects on time and in budget. This is where having a few options and evaluating these (while more time-consuming) can assist decision makers, increasing their confidence in the business case owner and assisting them in weighing up short- versus long-term investments to help achieve an objective or solve a problem.

If you want to read more about these financial aspects, I recommend reading *Financial Intelligence*,[11] which provides a straightforward yet detailed approach to understanding these topics further.

- **Implementation plan** – this should show an indicative road of activities and milestones for implementing the technology solution against a timeline. It should be detailed enough to face scrutiny, but it is a fine balance between being overwhelmingly detailed versus vague. The plan should show decision points as well because this provides an opportunity for the project to reset, replan or switch focus depending on how implementation is going and any changes happening within the organization which might impact the project. Within technology projects, you have different phases of work that will include activities outlined in Table 9.1.

A cautionary note to consider when developing the initial plan is in relation to the approach that the project will take. Engaging with vendors will help

Table 9.1 Activities that could be in a supply chain technology project

Phase	Examples of activities
Planning and analysis	• Identify the problem or opportunity that the technology aims to address • Gather data to understand the 'as is' and identify the requirements of users for 'to be' (needs assessment) • Identify suitable technology to address the 'to be' requirements, get demonstrations of solutions and evaluate the 'best fit' solution and vendor with indicative costs • Create the business case to justify the investment in the new technology • Create a project schedule with milestones and deadlines
Design	• Develop user stories and prioritization • Conduct design sessions with business and IT teams • Complete 'fit-gap'of requirements versus system proposed • Create a detailed design of the technology solution, including architecture and infrastructure
Build	• Build the technology solution, including coding, configuration and data integration, loading and validation • Create training documents and plan to train users • Test individual components of the solution to have confidence they work as expected
Test	• Test the solution so you're confident that all components work together seamlessly • Performance test solution • Test the solution in a controlled environment so it meets the requirements • Test the solution with users (user acceptance training [UAT]) so it meets their needs
Deployment	• Deploy the solution to the production environment • Configure the solution to meet the specific needs of the company • Provide training to users on the new technology • Go live with solution

(continued)

Table 9.1 (Continued)

Phase	Examples of activities
Support	• Monitor the performance of the solution so it aligns to expected outcomes • Gather feedback from user community to identify areas for improvement • Evaluate the ROI of the technology implementation and run lessons learnt sessions to capture ideas for future implementations • Provide ongoing support to users, including troubleshooting and maintenance • Perform regular updates so the solution remains secure and up to date • Continuously monitor the solution to identify areas for improvement

you understand the approaches other companies have taken with implementing similar solutions but be mindful that others may have done activities a specific way to either gain benefits early or had specific capability gaps that they needed to prioritize. Frequently, a vendor sunsetting a system means that companies have had to urgently find an alternative solution.

There are three common approaches taken for implementing technology projects: **waterfall**, **agile** or **hybrid** (see Figure 9.2).

Firstly, the waterfall approach is the traditional linear approach where projects are managed in phases and each phase is completed before commencing the next phase. It is suitable where the project requirements are defined and there are minimal changes expected.[12]

The agile approach sets up 'sprints' of activities and was born from having to manage projects at pace with consistently changing needs. It prioritizes continuous delivery and improvement for developing, delivering and sustaining complex products.[13]

The hybrid approach, as you can imagine, sits somewhere in between. It is used within the waterfall approach to bring speed to building and testing elements of a solution.

Over the last 20 years, companies have moved away from the waterfall approach to hybrid, primarily for significant long-term supply chain technology projects. Where the agile approach is primarily used is in the creation of one-off 'homemade' solutions, typically to solve a specific problem. In the example that follows this section, we will look at how the hybrid approach can be used effectively to drive incremental features on a minimum viable product that follows a waterfall approach.

EXPERT INSIGHTS

The agile approach works best in small, contained projects where the change is only within the function of supply chain, where the risk can be managed more effectively. When you go beyond the supply chain function, waterfall is a more ideal approach, as more assurance of knowing where you are currently and what you're aiming for is key when spending potentially millions of pounds.

The hybrid implementation approach is favoured if the technology will affect other functions. If your company proceeds with the agile approach, the importance of a well-orchestrated approach, detailed with all team members' roles agreed, provides clarity and predictability, especially when substantial investments are at stake.

When choosing the implementation approach, resource requirements, allocations and budgeting need to be considered. What time commitment do we require from our people? Do we need contractors, backfills etc? How will we manage allocations of time? Do we need more people from the business during the 'design' phase, fewer during the 'build' phase and then more again as we get nearer to 'test'? With this information, a bottom-up budget

Figure 9.2 Different approaches for implementation

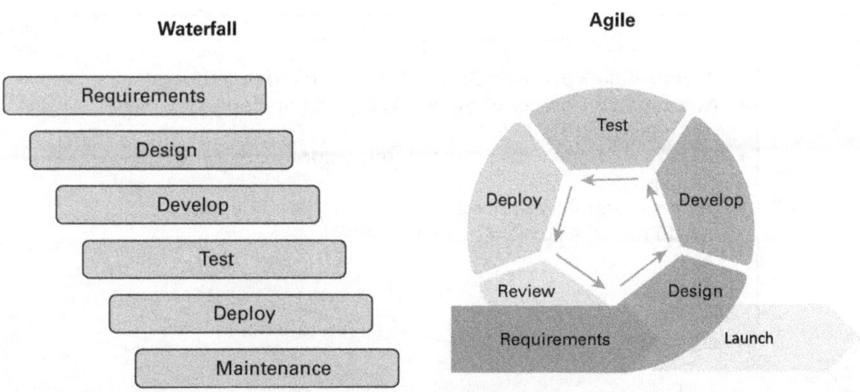

should be developed, outlining costs over time, which helps the finance team plan and also will show the impact of anything changing during delivery. Companies struggle with doing these activities because the decision makers may never have been involved in such a project before and it is an area that can change throughout the project. At the outset though, decisions have to be made on whether the company has enough of the right people to work on the project and therefore doesn't require much external support or whether roles need to be outsourced to a third party to support. In my experience, the knowledge business users bring is invaluable as they know the business and what the solution is solving, so they should be part of the team. At the same time, having external consultants can be valuable since they provide expertise the business may not have and the objectivity to challenge the company. For example, they will ensure the company is not just replacing the technology with limited revisions on the processes and operating model, since if that is the case, why are you investing? Technology solutions should be an enabler to improving, modernizing and/or standardizing processes and ways of working since they need to justify their existence and provide value to the company.

The acronym, RACI, is helpful at this stage, as it outlines the expectation on all people and parties who will be involved with the project[14] (defined in Figure 9.3).

Sometimes it is difficult to have only one party in each category due to the complexity of the activities and the interdependencies of different parties in achieving it but when that does occur, make sure the leader is stated.

Within the budget, the contingency fund should be agreed because the chances of optimism at the business case stage are typically much higher than when at the project delivery stage. The writer and economist Albert

Figure 9.3 RACI

- Responsible – people doing the work, completing the task
- Accountable – owner of the work, sign off and approve when task is complete
- Consult – people who provide input before work is completed
- Inform – people who need to be kept updated on progress but do not contribute directly to the task

Hirschman refers to it as 'the hiding hand', which is when project owners underestimate the problems likely to arise but also underestimate their capacity to address any problems that might occur.[15]

Risk management and mitigation strategies must be documented at this time as well and be maintained throughout the life of the project. One of the key misses at this stage can come with 'blindness' to getting the project approved at any cost without considering a raft of risks that are inherent to any change initiative. These should be brainstormed with various teams that are involved in delivery but also with the receivers of the solution. You won't be surprised to read that one retailer decided to go live with a new supply chain technology solution just before its peak period of sales – a fundamental miss by all parties that was never thought of as a risk until during user acceptance training (UAT), they were struggling to get users who had time to assist with the testing.

When developing a business case, review and iterate with colleagues in various functions so that as a team you can stand behind the business case. Use forums to challenge the case for the following (non-exhaustive) points:

- Lack of clear objectives.
- Insufficient data and metrics, with questionable assumptions and subjective information.
- Scope not ring-fenced and poorly defined.
- Optimistic resource allocation, assuming a colleague can take on additional responsibilities with little consideration for their other priorities.
- Obvious risk management points, narrow in focus.

A theme that has become popular over the last 10 years is to look at firstly developing a proof of concept (PoC) / pilot / minimum viable product (MVP) as almost the first part of approval for a business case. The 'prove it' phase of a PoC approach is a trend that has become common. You take a condensed version of the project and test it in a market, for example, to prove or disprove the business case but also to iron out any risks that may not have been considered when creating the original business case. This approach mitigates stakeholders from having to sign off significant investments based on a paper and some vendor demonstrations. Jeff Bezos at Amazon took this to extremes when deciding on whether to proceed with the Amazon Prime offering. He had a two-way door approach, where you can walk through the door and if you don't like what you see, you can walk back. His justification was that the time and cost spent debating could be more than just trying it and seeing.[16] So, in February 2005, it launched. It had been trialling other offers, which were complicated to execute with its technology architecture at the time, but one of its competitors was offering a deal on shipping, so it felt it needed to offer something to entice customers to buy more and more from it. Prime offered a flat fee for free shipping. The outcome was highly favourable and it has never looked back since.[17]

A PoC still requires going through the activities outlined previously to populate the case but seeks less financial funding and reduces risk by being done in parallel with business as usual. The purpose of it is to test the feasibility of a proposed solution, validate business and technical assumptions and demonstrate value to stakeholders. Plus, it gives a company the opportunity to evaluate several vendor solutions through a short exercise. Its advantages are that it can help reduce risk by seeing potential issues early before going into a full implementation. It can help reduce costs in the long term by identifying potential problems and issues early and reduce the likelihood of costly rework or abandonment later. It provides insight to refine and improve the solution design and give stakeholders confidence to buy into the larger request for funding of a full-scale implementation.

However, one drawback of a PoC is that it can lead to scope subjectivity, since do you cover enough of the capabilities provided by it to give a 'true and fair' view to base future decisions on? Or do you cover too many capabilities and it becomes a costly exercise? This can set unrealistic expectations on the solution's functionality, performance and benefits when scaled and is a challenge to integrate with existing systems, so the PoC is usually run independently from the current systems.

Whatever your approach and the company's way of funding supply chain management technology investments, following the outlined activities provides you with guide rails to produce a thorough proposal and should assist you with answering any stakeholder's questions and concerns.

Next is an example of bringing these elements together to build a business case for the implementation of a planning technology solution. It is not exhaustive and is based on many business cases I have been involved with in the planning domain.

EXAMPLE

Context

Company A is a multinational consumer products company that primarily sells its products through retail channels. It has a supply chain that operates in various countries, with manufacturing plants located in many of them. It has a sales presence in over 100 countries globally.

Business problem statement

Our current planning system is nearing its end of life. Therefore, we need to invest in a new planning system that meets current and future requirements (to be determined during the design phase) for supply chain planning. We have more than 200 users who cover various roles of planning (demand planning, integrated business planning, supply planning, production planning and distribution requirements planning) based in our main planning hubs in five regions. They handle a portfolio of 10,000 finished goods made up of 1,000 raw materials and 5,000 packaging materials.

This project presents an opportunity for us to streamline interactions between a variety of systems and clean up our data.

Solution overview

A planning system enables us to optimize planning across our supply chain and provides a centralized platform for data collection, analysis and reporting. The system will assist users in making informed decisions in near real time with the specific capabilities listed in Figure 9.4. It will provide us with the ability to address use cases that are business-critical or cannot be completed today due to system constraints or are time-consuming because of how many of the activities require to be completed manually 'out of the systems'.

In this section, documenting critical requirements of the solution must be worked through with the chosen independent software vendor (ISV).

Table 9.2 shows a few examples of use cases you might consider. These should be focused on the activities that the supply chain planners are unable to do today or take many hours of preparation. Ideation sessions using unconstrained thinking would benefit the planners at this stage to devise differentiating capabilities. The objective is to get the users engaged and help them to focus on the proactive elements of their role, which many planning systems and implementations have failed to achieve to date.

Figure 9.4 Example of 'planning' system overview of capabilities

Demand planning	Supply planning	Integrated business planning	Production planning
• Historical data analysis – analyze historical sales data to identify trends and patterns • Near real time data analysis – analyze real time data to identify changes in demand patterns • Promotions and advertising – understand effect promotions and advertising have • Product placement – understand effect product placement strategies have on influencing demand	• Supply and demand matching – match supply with demand for optimal allocation • Reorder point calculation – calculate reorder points to maintain optimal inventory levels • Inventory analysis – analyze inventory levels to identify opportunities for optimization **Optimization** • Linear programming – optimize production plans and schedules • Mixed integer programming – optimize production plans and schedules • Heuristics – use heuristics to optimize production plans and schedules	• Production planning – plan production to meet demand • Inventory planning – plan inventory to meet demand • Supply chain planning – plan supply chain operations to meet demand • Demand sensing – sense changes in demand patterns to inform business decisions • Demand – shape demand to meet business objectives **Distribution requirements Planning** • Flow analysis – analyze flow through the distribution network • Inventory analysis – analyze inventory levels	• Material requirements planning – determine material requirements for production • Capacity planning – plan for sufficient capacity to meet production requirements • Production scheduling – create production schedules to meet demand **Analytics** • Performance metrics – track and assess key performance metrics, such as throughput and lead time • Root cause analysis – identify and address root causes of production issues • Predictive analytics – deploy advanced analytics to predict production performance and opportunities

Scheduling	Simulation	
• Finite capacity scheduling – schedule production based on available capacity • Infinite capacity scheduling – schedule production without considering capacity constraints • Just-in-time scheduling – schedule production to meet just-in-time demand	• Discrete event simulation – simulate production processes to analyze and optimize performance • Continuous simulation – simulate production processes to analyze and optimize performance	• Reorder point calculation – calculate reorder points to maintain optimal finished goods inventory levels • Inventory classification – classify inventory to prioritise optimization efforts • Load optimisation – optimize loads to reduce transportation costs

Figure 9.5 Illustration of system landscape

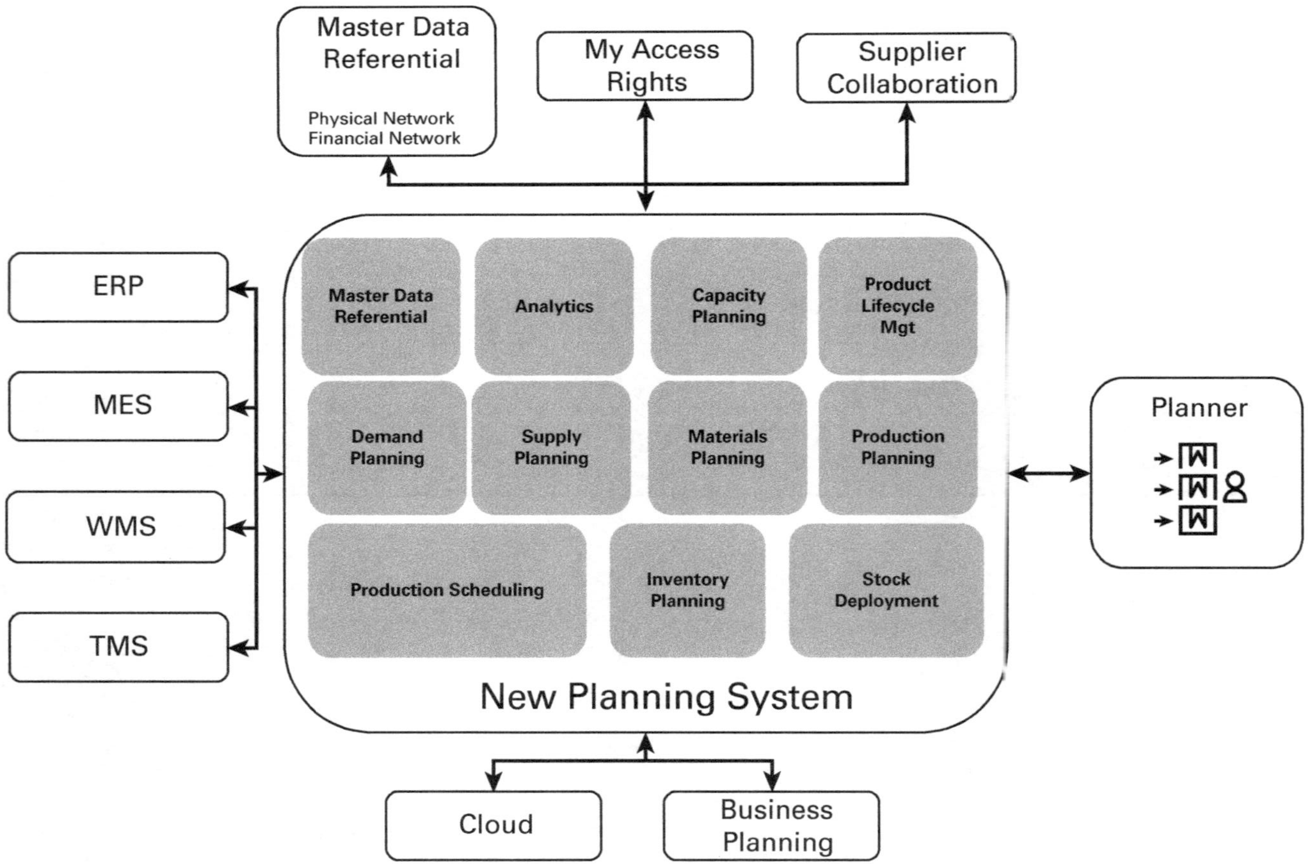

Table 9.2 Example use cases aligned to different planning modules

Demand planning	Supply planning	Integrated business planning	Production planning
Model selection and segmentation	Minimize inventory obsolescence	Planning-based alerts	Scheduling unplanned events in short term
Forecasting output from system should not be a 'black box'. Planners must understand how the forecast was created and enable change model parameters	Planner needs to be able to model revenue and cost impact of phase in / out of portfolio of SKUs, looking at it from a variety of scenarios such as over short vs. mid to long	Planner needs to be able to see when inventory costs are exceeding the minimum / maximum cover set as a business rule, guide the user to the bottleneck and provide recommendations and financial impact	In response to an unplanned event, the planner needs to be able to update the schedule without having to recreate a new schedule of activities after they have reviewed recommended scenarios provided by solution

Furthermore, system context is shown in Figure 9.4, with the capabilities of a new demand planning system aligned to legacy architecture.

Figure 9.5 shows nine systems that will interface with the new planning system in this example. Therefore, determining the number of interface points (i.e. the connections required, see Chapter 7 for more details) is critical. It is also critical to determine how data will move between the systems, whether in near real time, done on demand when a user makes a specific request or done in a batch approach such as being updated overnight. Knowing this information will also inform integration time and costs.

Table 9.3 outlines the 71 interfaces and the data that will need to move between the planning system and other systems as an example.

Table 9.3 Data and interfaces

Kind of data	Type of data	Target
Master data	Administration	2
Master data	Demand	2
Master data	Distribution	2
Master data	Inventory	-

(continued)

Table 9.3 (Continued)

Kind of data	Type of data	Target
Master data	Network	3
Master data	Planning	1
Master data	Product	8
Master data	Production	8
Master data	Supply	5
Master data	Other	5
Transactional data	Demand	7
Transactional data	Distribution	5
Transactional data	Financial	4
Transactional data	Inventory	4
Transactional data	Order	4
Transactional data	Production	4
Transactional data	Supply	4
Transactional data	Other	3

Benefits analysis

Benefits for this type of project can be far-reaching and in Chapter 10, we will explore a variety of levers available to determine a realistic aspiration (benchmark) to target post the solution implementation. It is difficult to put credible numbers to this due to finances being company-specific, but the expectation would be to see improvements in several areas, such as forecast accuracy, which then reduces lost sales, inventory costs and waste. There could be a reduction in supply chain planners required to do the roles, as the technology makes the activities faster. From a supply perspective, having more confidence in plans could improve volume commitments and therefore pricing. From a transportation perspective, more certainty on plans could reduce the number of one-off deliveries required and reduce the need to use expensive forms of transport, such as air.

In some companies, benefits are derived from a target linked to revenue. However, I believe that more consideration should be given at this stage to the levers that realistically will be improved by implementing new (or modern) capabilities.

Cost analysis

The indicative cost to implement the planning system solution globally is made up of the elements outlined in Table 9.4. Once a chosen independent software vendor (ISV) is agreed, both one-off and annual costs can be firmed up.

Assumptions could include:

- The implementation cost for ISV will be finalized when the preferred ISV is selected.
- System integrator costs are estimated based on similar projects.
- Business integrator costs will be agreed aligned to roles the company will assign internally and externally.
- Internal project resource costs are assumed as backfill for current roles.
- SaaS licence costs will be finalized when the preferred ISV is selected and the number of users of the solution is agreed.
- Running costs include maintenance and regular updates to the system.

Table 9.4 Indicative cost areas

Costs
Implementation cost – ISV
Implementation cost – system integrator
Implementation cost – business integrator
Internal project resources
Contingency fund for implementation
SaaS licence
Data security
Ongoing system enhancements
Maintenance

At this stage, the company most likely will be working with several vendors to understand a match to its unique requirements and may not have engaged integrators (system and business). The assignment of roles between the company, external consultants and contractors can vary at different phases of the project. For example, during the design phase, the need for business consultants' expertise could be focused on processes and then during the build phase, they could be more focused on organizational elements such as change.

Financial analysis

Once benefits and costs are known, the financial formulas outlined previously in the chapter can be determined. This should be completed in conjunction with your finance department.

FINANCE'S 'OUTSIDE IN' PERSPECTIVE

We have touched on the role of the finance function and engaging them early and frequently through iterations of the business case several times in this chapter. With that in mind, this is a deep dive into further things to consider as you develop the business case, supported by insights offered by a finance expert.

There are several elements your finance team will be interested in understanding and supporting you with. Alignment with the business goals seems obvious but it is imperative to test that what is being proposed aligns with the overall strategic objectives of the organization. From supporting long-term goals to enhancing competitive advantage and maintaining the current operations asset base, the need and priority of the initiatives must make sense when looked at with all the other competing investment requests versus the cash flow available. The initiatives should not be a surprise to the leadership team and will usually be on a roadmap of priorities that are discussed regularly. Decisions on investments made are evaluated against a multitude of factors with less strategic initiatives moving up and down the priority list regularly based on how the business is performing, its cash flow situation, the external economic climate and the leadership's view at the time on what will provide them with competitive advantage or improve impact on customers, for example.

The types of activities we would expect finance to do will also be a key part of the case, as already outlined. Conducting cost-benefit analysis to evaluate the financial impact of the investment, from assessing initial costs, ongoing maintenance expenses, costs to sunset legacy systems and on the other side of the equation, the potential savings and/or revenue generation is crucial.

There are standard financial calculations (e.g. ROI, NPV, payback) that the investment in most instances will need to pass, with the only exceptions being if a system fails or if continuing with it leads to increased risk to the organization or an innovation that makes sense to quickly approve. Benefits should be simple and measurable and point to levers of improvement such as efficiency, reduced operational costs and enhanced customer satisfaction, to name a few. Your finance business partner should maintain an objective view of how the investment supports the strategy and economic health of the organization. They should seek a stretching but realistic scenario as the base case for investments and perform sensitivity analyses around it. The finance role involves challenging biases from different perspectives, ensuring realistic projections and sometimes encouraging a wider scope of business case perspective and more ambitious goals. They must balance financial hurdles with strategic priorities, supporting the supply chain owner in presenting a well-rounded case, even if the payback period is long. Ultimately, it's about the leadership being able to make decisions based on transparent and realistic information, discussing risks and opportunities openly and ensuring decisions align with the overall strategy.

On risk management, perspective will be required to identify and mitigate risks associated with the implementation, such as cybersecurity threats, compliance issues and any potential disruptions to existing operations. One watch-out which is subjective but needs to be considered is the scalability of the technology and its ability to accommodate future growth and any changes known that could occur within the supply chain due to new directions in business strategy (e.g. M&A, carve outs, exiting / expanding into new markets). There is a high chance some of these may not be known to the supply chain team creating the business case.

It is easy for supply chain leaders to have a narrow perspective, fixating on getting the business case perfect. While we would all like to think we weren't biased, bringing in objectivity to support us is crucial from the outset. Within all organizations, the gatekeeper of funds to support initiatives from a capital perspective sits with the CFO and their team. Most supply chain technology investments will be over clip levels that most senior non-finance team members cannot approve (i.e. there will be limits of spend that depend on the organization and are usually explicitly stated and controlled). Therefore, working collaboratively with them and other relevant functions (e.g. commercial), where appropriate, that gain from the investment too will be crucial. This is to help develop the case but also importantly, have them as advocates and supporters of the investment from as early in the process as possible. With that in mind, the CFO is also most likely the person who needs to be convinced the most because they will then help the board understand the investment and make their recommendations. The main reason for this is trust, so other decision makers know a business case has been rigorously analysed and developed when the CFO and the finance team stand behind the initiative.

One final thought is the business case: the depth of the case must be nuanced, according to the culture of the organization. Many companies that operate globally and are publicly listed will typically have a formal process compared to private and smaller companies, which might not be so formal with their templates and documentation. Either way, creating a robust business case requires supply chain and finance to be joined at the hip, supporting each other to get to the right outcome.

Implementation plan

Many days can be spent on the plan and determining the suitable approach that links to the company's ambitions and ability to handle change. Typically, necessity drives priorities or an interdependency on another programme (e.g. ERP upgrade). The following figures show high-level (Figure 9.6) and mid-level (Figure 9.7) plans to give you an idea of some of the key messages that need to be shared at this stage.

How much detail you need to show should be agreed with collaborators and provided with supporting assumptions.

Figure 9.8 shows a plan of stages for 'demand planning' pilot and rollout. Some assumptions for this could include:

- The objective is to achieve a fast rollout to minimize time to capture benefits.
- Preparation for the rollout will commence towards the end of the 'pilot' phase.

Figure 9.6 Example of high-level plan

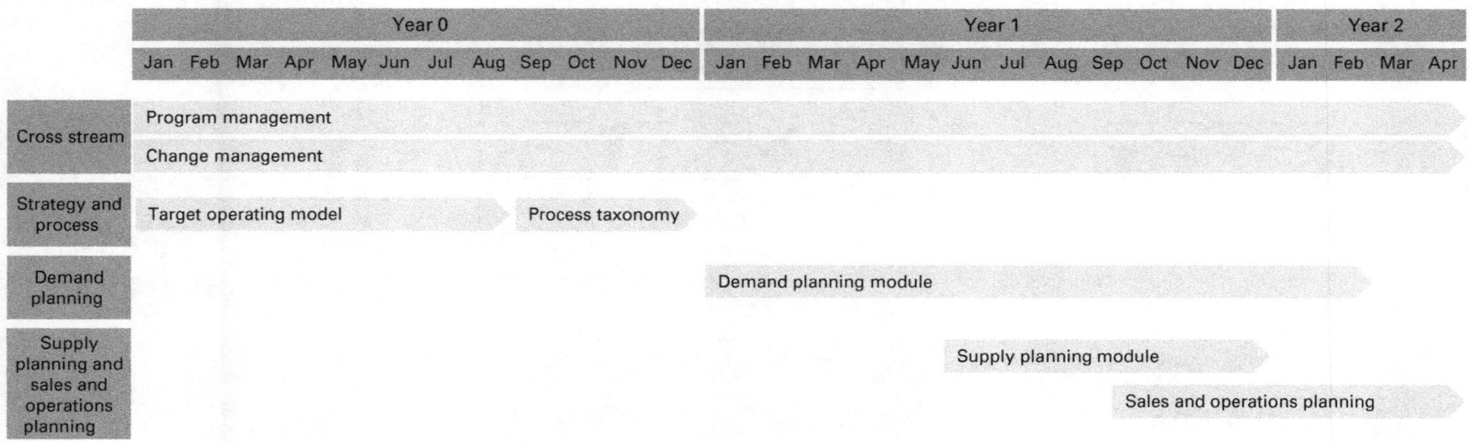

Figure 9.7 Example of mid-level plan

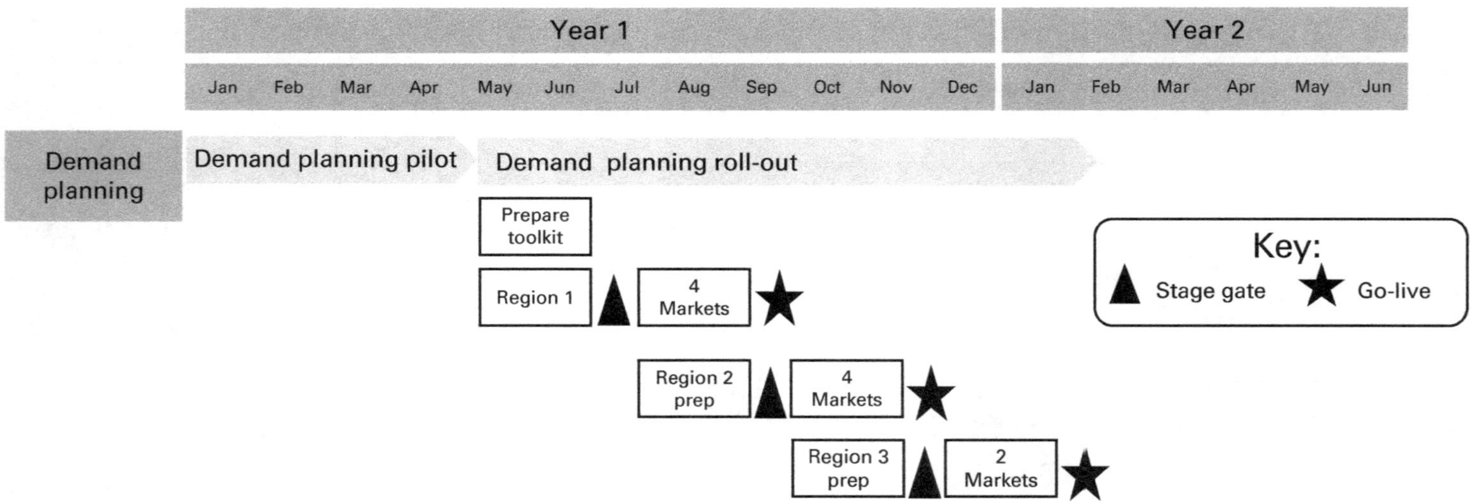

Figure 9.8 Example of demand planning pilot plan

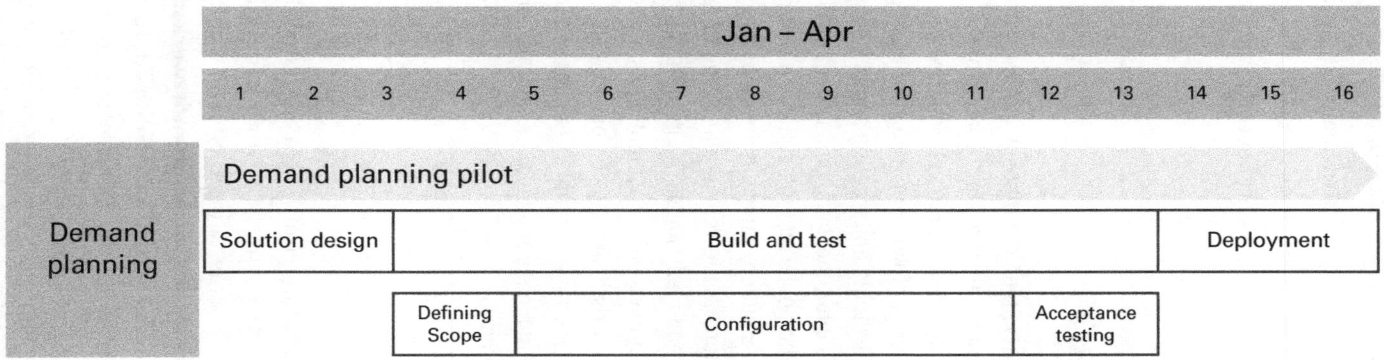

- 'Demand planning' and 'supply planning' can be executed concurrently during the rollout due to different company users being engaged.
- The plans may be affected by another major IT project, which, in current planning, is not considered to be happening, but requires impact assessment.
- Enhancements are handled in a parallel track but delivered only after the rollout.

EXPERT INSIGHTS

When reflecting on past implementations, a common theme appears with projects that did not go to plan: the over-promise of technology. While on paper, it may make sense, you need to ask whether you are using the technology to be pioneers or followers. If you are hoping to pioneer, extra caution is required. You should be targeting a short, sharp 'pilot' of functionality and that is what the initial business case should focus on achieving, which can then build into full-scale implementation. If you are implementing a proven solution, speaking to customers should be done but don't just take the customers the vendor wants you to speak with. Ask to speak with those who had challenges. The information gathered at this time will be beneficial for planning and documenting as a potential future problem. Also don't underestimate the complexity and risks associated with undertaking initiatives because in many cases, these solutions from day one will be exposed to business-as-usual conditions, which will have a direct impact on customers.

Defining RACI

Next is determining resource requirements and defining roles and responsibilities (RACI) on the project. There are a variety of ways to approach showing resource requirements. A detailed overarching spreadsheet showing effort against activities for each relevant workstream in conjunction with ISV, SI and BI providers could be developed, for example. Whichever approach is used, when showing resource requirements to stakeholders, visual approaches always trump. Provide a commentary or be ready to explain why changes occur in resourcing, which align to effort requirements at different phases of a project life cycle. Figure 9.9 and Figure 9.10 show two different ways that you could present a resource profile.

Figure 9.9 Resources profile bar graph

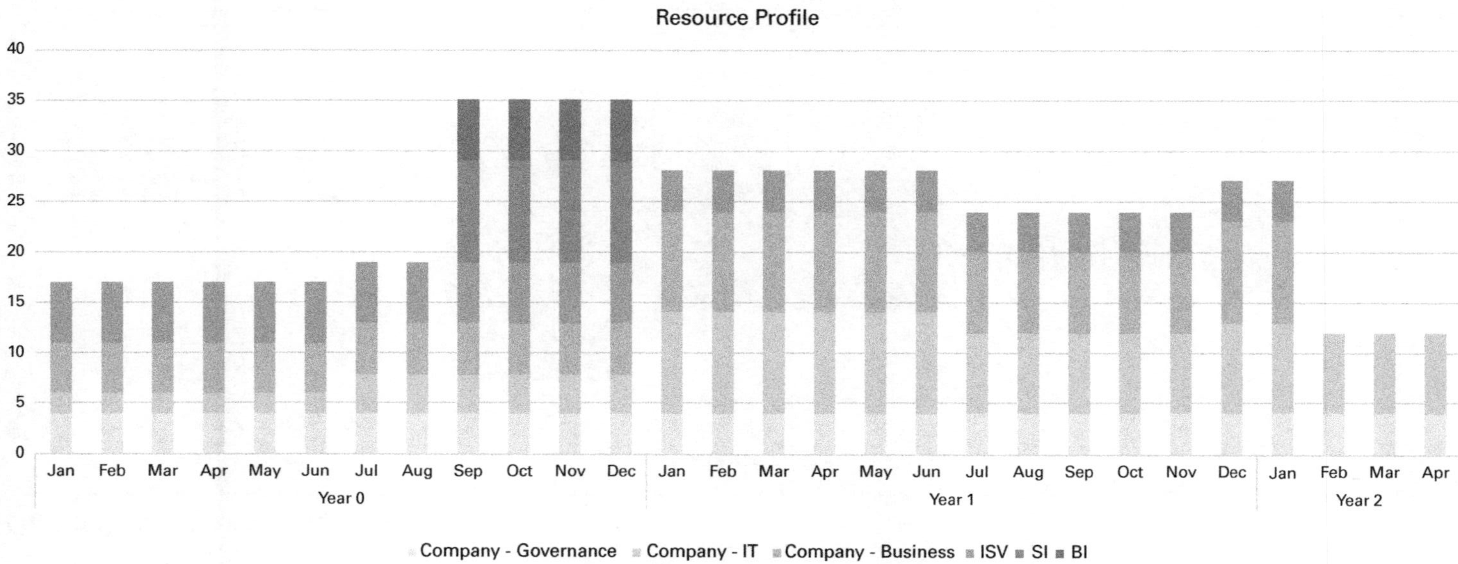

Figure 9.10 Resources profile stack graph

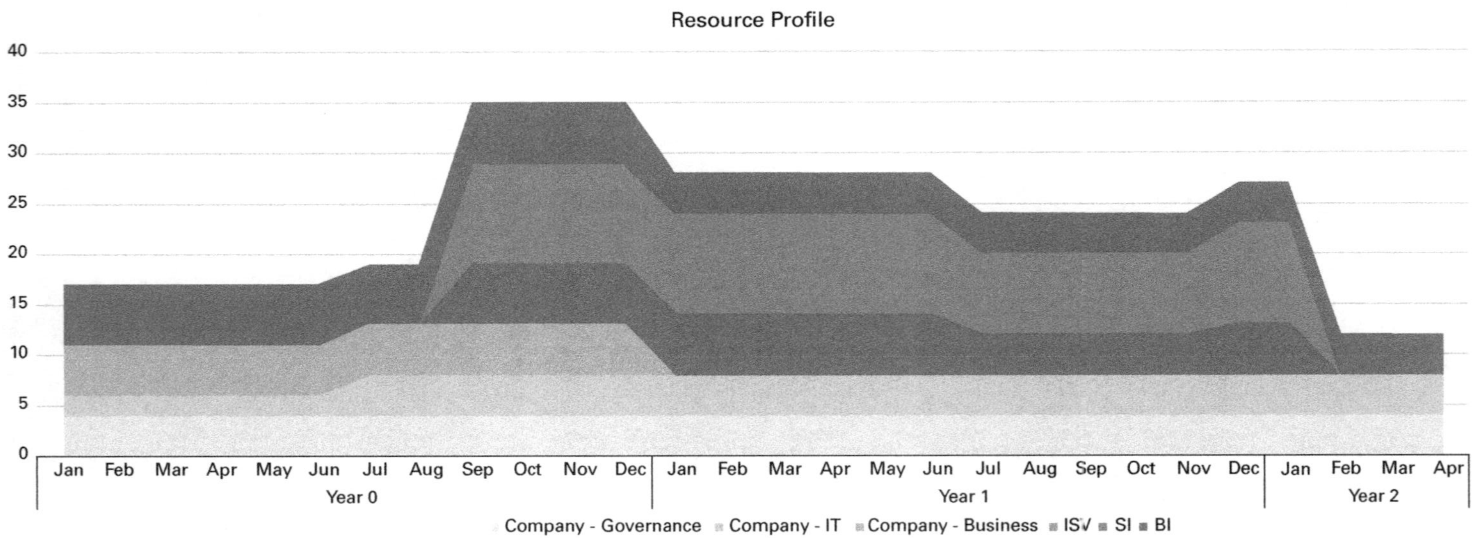

Figure 9.11a Detailed RACI example across all phases (not exhaustive)

Sub-phase	Activity	BUS	ISV	SI	BI	Group	Deliverables
Project planning	Manage project plans and resources	A	C	R	R	PM / PMO	Project plan, resource capacity plan
Project planning	Manage internal ISV plans	-	A, R	-	-	PM / PMO	
Project planning	Manage internal SI and BI plans	-	-	A, R	A, R	PM / PMO	
Monitoring	Manage the steering committees	A	C	C	R	PM / PMO	Steering committee presentation deck
Monitoring	Set up and run the design authority meetings	A	C	R	C	PM / PMO	Design authority presentation deck
Monitoring	Set up and run the project status meetings	C	C	C	A, R	PM / PMO	Project status presentation deck
Monitoring	Set up and run the project risks matrix	C	C	C	A, R	PM / PMO	Project risk matrix
Monitoring	Set up and run change requests	A, R	R	R	R	PM / PMO	
Monitoring	Conduct the project quality assurance reviews	C	C	A, R	A, R	PM / PMO	
Project preparation	Prepare and implement project management document templates (governance instances)	C	C	A, R	A, R	PM / PMO	Project management templates

(continued)

Figure 9.11a (Continued)

Sub-phase	Activity	BUS	ISV	SI	BI	Group	Deliverables
Project preparation	Set up project management tools (activity tracking, issue tracking, risk tracking...)	A, R	C	I	R	PM / PMO	
Project preparation	Prepare project management dashboards (KPIs, status, issues, risks...)	C	C	C	A, R	PM / PMO	
Project preparation	Prepare project quality assurance plan	C	I	I	A, R	PM / PMO	Quality plan
Project preparation	Launch governance instances and logistics (participants, agendas, invitations...)	A, R	C	I	-	PM / PMO	
Project preparation	Prepare onboarding support toolkit for project team	A, R	C	I	-	PM / PMO	
Project planning	Execute communication plan	A	C	C	R	Change mgt	Communication plan
Change management	Measure solution adoption	A	C	C	R	Change mgt	

Figure 9.11b RACI example for pilot phase (not exhaustive)

Sub-phase	Activity	BUS	ISV	SI	BI	Group	Deliverables
Project launch	Organise site visits, meetings and interviews with business	A, R	C	C	-	PM / PMO	
Project launch	Plan and schedule solution activities for PoC: Sprints	C	A, R	C	-	PM / PMO	
Project launch	Plan and schedule meeting cadence for project management	A, R	C	C	R	PM / PMO	
Project launch	Prepare and conduct kick-off	A, R	I	C	R	PM / PMO	Kick-off deck
As-Is	Collect and share documentation of the As-Is	A, R	C	C	-	Strategy	
As-Is	Analyse As-Is documentation	I	C	A, R	A, R	Strategy	
As-Is	Conduct interviews of selected operational stakeholders	I	C	A, R	A, R	Strategy	
As-Is	Conduct work shadowing of key operational stakeholders	I	C	A, R	A, R	Strategy	
As-Is	Develop high-level process taxonomy of the As-Is	I	C	A, R	A, R	Strategy	
As-Is	Develop critical observations: pain points, areas for improvement	I	C	A, R	A, R	Strategy	

(continued)

Figure 9.11b (Continued)

Sub-phase	Activity	BUS	ISV	SI	BI	Group	Deliverables
Maturity assessment	Define maturity assessment approach	C	C	A, R	A, R	Strategy	
Maturity assessment	Prepare maturity assessment materials	C	C	A, R	A, R	Strategy	
Maturity assessment	Assess maturity level (self-assessment)	A, R	I	I		Strategy	Maturity assessment report
Maturity assessment	Assess maturity level	I	C	A, R	A, R	Strategy	Maturity assessment report
Maturity assessment	Present maturity assessment outcomes	C	C	A, R	A, R	Strategy	
Strategy, advisory	Define vision narrative	A	C	R	R	Strategy	
Strategy, advisory	Define target operating model	A	C	R	R	Strategy	Target operating model
Strategy, advisory	Develop target operating model roadmap	C	C	A, R	A, R	Strategy	Recommendations on transformation path
Strategy, advisory	Conclude and document business requirements	C	C	A, R	A, R	Strategy	Business requirements
Solutioning	Set up hosting / cloud	I	A, R	I	I	Integration / data	
Solutioning	Implement namespace in ERP	A, R	C	I	I	Integration / data	
Solutioning	Set up basic planning cycle in the ISV solution	I	A, R	I	I	Functional	ISV solution with basic configuration

Solutioning	Set up basic integration with the ISV solution	C	A, R	I	I	Integration / data	ISV solution with basic integration
Solutioning	Provide a data sample (in ISV format)	A, R	C	I	I	Integration / Data	
Solutioning	Design solution: Critical use cases	C	A, R	I	I	Functional	
Solutioning	Configure critical use cases	I	A, R	I	I	Functional	
Solutioning	Prepare and present the solution and processes to the project team	I	A, R	I	I	Integration / data	
Solutioning	Validate, provide feedback	A, R	C	I	C	Functional	
Solutioning	Go / no-go decision (for critical use cases)	A, R	C	I	C	Functional	
Solutioning	Identify key users	A, R	C	C	I	Change mgt	
Change management	Initial change leadership alignment and vision	A	C	C	R	Change mgt	Change management strategy
Change management	Perform initial global stakeholder analysis	I	C	C	A, R	Change mgt	Stakeholder map
Change management	Initial global high-level change impact analysis	A	C	C	R	Change mgt	Global high-level impact analysis
Change management	Develop initial change and training strategy, approach and tools	A	C	C	R	Change mgt	Training strategy, approach, tools
Project planning	PoC exit criteria met, signed off, go / no-go decision	A, R	C	C	C	PM / PMO	

When creating a RACI, it is critical that it is documented and agreed with relevant stakeholders internally and externally prior to beginning the project. Ambiguity on who is accountable and responsible, especially if more than one party is assigned to a specific activity, can become a challenge if something does not go to plan or a change is required. Figure 9.11a and Figure 9.11b provide a snapshot of relevant activities within each phase and RACI.

Table 9.5 provides examples that make up the business case for a supply chain planning technology solution. There are many details to be worked through and parts of the case will be more informed and detailed than other elements. Also, decisions aligned to scope and RACI will determine the effort from the various parties you engage to support your company in completing the project. It is important to keep this document live and

Table 9.5 Risks and their mitigation strategies examples

Risks	Mitigation strategies
Data quality	Assess data quality, based on audit, organize and agree the fixing of data quality issues in source systems
Historic data to be migrated	Agree how much historical data needs to be migrated
Data for testing	Document data preparation activities starting during PoC
Availability of stakeholders with knowledge of data in integrated systems	Organize the involvement of stakeholders starting from 'target operating model' phase
Number of interfaces to be developed	Reuse as much as possible interfaces for extractions and assess the feasibility / relevancy of interface development
Complexity of running simultaneously in production existing and target system	Agree in 'PoC' activities to build the cutover plans to minimize effort and risk in transitioning to new system
Technical complexity of cutover	Assess the feasibility to split the complex cutovers in a set of more simple cutovers (anticipation of some moves to production)

updated as the team goes through the project. As Clay Shirky, Vice Provost for AI and Technology in Education at NYU, suggests to his students, doing post-mortems mid-project is a good thing. People often can see what will go wrong as the project progresses, so the sooner they are heard, the better.[18] This is very applicable to going through a supply chain technology implementation. However, I would suggest that this be done at specific milestones or when nearing the end of each phase.

Conclusion

Developing a business case for supply chain management technology investments is a demanding but rewarding activity, allowing you to evaluate potential benefits and costs of a new solution. When completed by following the approach outlined in this chapter, it can help you prepare thoroughly in advance before moving into its execution. By following these practices, you will develop a comprehensive and robust case that will gain stakeholders' trust and understanding to appreciate and support informed decisions. Engage stakeholders frequently throughout the development of the case, use data and metrics, evaluate several scenarios and create a consistent and clear narrative assisted with visual aids to help communicate complex information to secure funding.

It should be noted that although you have been successful in getting approval, reiteration of plans and the consequences of that on the timing of benefits and costs will continue to be an activity that needs to be undertaken. No project that I have been part of doesn't have some change in plans; keeping track of changes and discussing and agreeing on these should be part of the day-to-day activities throughout the life of the project.

When you look at the business case, the benefits analysis, cost analysis, financial analysis, implementation plan and RACI should form the backbone of thinking and discussion to support making well-informed decisions in relation to supply chain technology solution investments.

While following what has been laid out in this chapter does not guarantee success (and it would be foolish to suggest that it would alone), it will provide you with a strong foundation to move forward on. No one goes into implementing any project expecting the worst, which is why it is important to flush out all potential outcomes as early as possible. Get comfortable with a level of risk and test whether the company stakeholders understand the risks and are supportive of moving forward.

Further reading

Berman, K, Knight, J and Case, J (2013) *Financial Intelligence* (Revised edition), Harvard Business Review Press

Endnotes

1 De Mello, A (1997) *Awareness*, Fount

2 Rider, B (2024) Algorithms are the best way to improve computing power, WordPress, https://wjrider.wordpress.com/2024/08/14/algorithms-are-the-best-way-to-improve-computing-power (archived at https://perma.cc/4C3G-7NYN)

3 Bottke, T, Kearns Manolatos, D and Troilo, G (2023) Do you really know the financial impacts of your digital transformation? *Harvard Business Review*, https://hbr.org/sponsored/2023/04/do-you-really-know-the-financial-impacts-of-your-digital-transformation (archived at https://perma.cc/SY8T-XFRC)

4 Bloch, M, Blumberg, S and Laartz, J (2013) Delivering large-scale IT projects on time, on budget, and on value, McKinsey, www.mckinsey.com/~/media/McKinsey/dotcom/client_service/Corporate%20Finance/MoF/PDF%20issues/PDFs%20Issue%2045/Final/MoF45_LargeScaleIT.ashx#:~:text=However%2C%2017%20percent%20of%20IT,under%20a%20normal%20distri%20bution (archived at https://perma.cc/73UX-GBDN)

5 Direct Commerce Central (2024) Migration to New Wincanton DC causes White Company angst, Direct Commerce Central, https://homeofdirectcommerce.com/news/migration-to-new-wincanton-dc-causes-white-company-angst (archived at https://perma.cc/9RPC-MK6S)

6 Morgan, A (2025) End Clothing plunges £43m into the red after warehouse troubles, *Retail Gazette*, www.retailgazette.co.uk/blog/2025/01/end-clothing-43m-loss (archived at https://perma.cc/NZA9-RA3B)

7 Ashcroft, S (2021) Top 10: Worst supply chain disasters in history, https://supplychaindigital.com/supply-chain-risk-management/top-10-worst-supply-chain-disasters-history (archived at https://perma.cc/4Q2G-XJHJ)

8 Flyvberg, B and Budzier, A (2011) Why your IT project may be riskier than you think, *Harvard Business Review*, https://hbr.org/2011/09/why-your-it-project-may-be-riskier-than-you-think (archived at https://perma.cc/E5BL-BZPF)

9 Unglesbee, B (2024) Nestlé hit by supply chain constraints in health business, Supply Chain Dive, www.supplychaindive.com/news/nestle-supply-constraints-health-science-IT-implementation/708503 (archived at https://perma.cc/7P9P-GPV3)

10 Bourne, M (2025) *APM Body of Knowledge*, Association for Project Management 8th edition

11 Berman, K, Knight, J and Case, J (2013) *Financial Intelligence* (Revised edition), Harvard Business Review Press

12 Royce, W W (1970) Managing The Development Of Large Software Systems, The Institute of Electrical and Electronics Engineers, IEEE WESCON, August 1970, pp 1–9, https://blog.jbrains.ca/assets/articles/royce1970.pdf (archived at https://perma.cc/S5HB-69X6)

13 Atlassian (2025) What is the agile methodology? Atlassian, www.atlassian.com/agile#:~:text=Agile%20is%20an%20approach%20that,Use%20free%20Agile%20roadmap%20template (archived at https://perma.cc/KFQ8-HC2N)

14 Kantor, B (2025) The RACI Matrix: Your Blueprint For Project Success, CIO, www.cio.com/article/287088/project-management-how-to-design-a-successful-raci-project-plan.html (archived at https://perma.cc/9WW9-4UDN)

15 Hirschman, A O (2014) *Development Projects Observed*, Brookings Institution Press

16 Amazon (2016) 2015 Letter to shareholders, Amazon, https://s2.q4cdn.com/299287126/files/doc_financials/annual/2015-Letter-to-Shareholders.PDF (archived at https://perma.cc/6C2S-F67A)

17 Del Rey, J (2019) The making of Amazon Prime, the internet's most successful and devastating membership program, Vox Media, www.vox.com/recode/2019/5/3/18511544/amazon-prime-oral-history-jeff-bezos-one-day-shipping (archived at https://perma.cc/Z7PD-DXSC)

18 Mstem (2013) Clay Shirky: Planning shan't trump reality, Civic MIT, https://civic.mit.edu/index.html%3Fp=1885.html (archived at https://perma.cc/JX42-7N3C)

Measuring success and ROI in supply chain technology investments

10

This chapter will discuss the importance of measuring success and return on investment (ROI) in supply chain technology investments, including key performance indicators and metrics.

CHAPTER AIMS

- Learn how to measure success and ROI in supply chain technology investments.
- Understand which systems drive benefits in today's supply chain.
- Know the key performance indicators and metrics for measuring success and ROI.

'Most business is probabilistic but everybody in business wants to prove and pretend that it's deterministic.'

RORY SUTHERLAND[1]

Having looked at the development of the key elements that make up a supply chain technology investment and at examples of challenges that need to be overcome at the outset of and during implementation, we now turn to the critical point of how to measure the success of an implementation, discerning what difference the new solutions have made to benefit the company.

EXPERT INSIGHTS

I drew on the experience of another expert in the field to provide the reader with further considerations when measuring the success of technology investments.

First should be the recognition that any investment in supply chain technology must align seamlessly with the overarching business strategy. This alignment is not merely a strategic choice but a necessity, ensuring that technology investments support business objectives and deliver tangible value. The process starts with the development of a multi-stakeholder map. This map is a comprehensive tool designed to understand the interests of various parties involved, including consumers, the board and other key stakeholders. By mapping out these interests, organizations can ensure that their technology investments are not only strategic but also inclusive, addressing the needs and expectations of all involved.

Evaluating technology investments needs to be a meticulous process. It requires executive alignment and the support of key players within the organization. Building a business case for these investments can be a time-consuming endeavour, sometimes taking up to a year. However, this thorough approach is crucial for the success of any project. It ensures that all aspects are considered and the investment is justified from both a strategic and financial perspective. Also, there are challenges to credibility with vendor solutions: do they have the stories, the scale in the solution for a global business? If they don't, then they need to focus on a 'local' market and get sponsorship through a pilot to trial and prove it. This raises a valid challenge on investments: what is the right approach – 'global' versus 'local'? Evaluating technology investments at a 'global' level, you are trying to treat all 'local' as the same; however, something that works well for one country may be far too complex for another. For example, I was involved in the implementation of a customer relationship management system where the solution developed made sense for a market with over 200 customers, however, when the solution was rolled out to a country which had 10 customers, the solution was a burden to them because it involved having to do many activities in the system that were not applicable. Experience suggests getting markets of scale onto similar a system and challenging whether other markets can go a different way as long as the overarching objective can be achieved and that outputs from the systems can achieve similar outcomes. So the real question becomes, does a one-size-fits-all global solution really provide the value to an organization going forward since very few actually deliver on what they say?

Return on investment (ROI) plays a significant role in decision-making. Typically, organizations aim for capital investments to have a payback period of three to five years. However, achieving this target can be challenging, especially for technology investments. These investments often require a more nuanced approach to identifying real savings and understanding their long-term impact on the business. Measuring the success of technology implementations involves navigating learning curves and managing delays. Many projects tend to be delayed and do not meet their initial timelines. Strong leadership is essential to navigate these challenges and effectively manage change. Regular surveys and feedback mechanisms should be considered to track user adoption and engagement with new systems. This approach helps identify areas for improvement and keeps users invested in the technology. Assessing the financial impact of technology investments is a complex task. While tracking financial returns is essential, attributing specific improvements to technology can be challenging due to multiple influencing factors. A holistic view is necessary to understand the overall impact and ensure that the investment is delivering the desired outcomes.

Compliance is another critical driver of technology investments. In certain industries, such as pharmaceuticals, compliance with regulatory requirements can necessitate significant technology investments. For example, the implementation of electronic batch records is driven by compliance needs. However, even in these cases, it is essential to seek financial benefits and ensure that the investment yields quantifiable returns.

In conclusion, the integration of technology in supply chain management is a multifaceted journey. It requires strategic alignment, thorough evaluation and strong leadership. By understanding the interests of all stakeholders, seeking financial benefits even in compliance-driven projects and ensuring scalability, organizations can successfully navigate the complexities of technology implementation. This journey is not just about the technology itself but about the value it brings to the business and the lives it touches along the way.

Measuring the success of supply chain technology investments is crucial to ensure that these investments result in the expected returns both from tangible (e.g. financial benefits) and intangible (e.g. improved user satisfaction) perspectives. There is a variety of metrics and practices that should be used to help you measure success.

Metrics

Metrics (some of which we have referred to in Chapter 9) can be cold and narrow in view and don't always provide a full picture of the benefits the company will gain. For example, there might be a significant change in how a business performs from the time that it decides to proceed with implementing a capability to completion and measuring how successful it has been. These types of metrics include return on investment (ROI), total cost of ownership and the payback period. ROI compares perceived benefits in financial terms versus the costs to realize the new solution. Total cost of ownership goes beyond ROI by including the costs of maintaining the technology.

Another challenge is that new benefits which may be indirectly linked to the new technology capability might not be considered, such as investment for growth. This is because during business case preparation, basing decisions on unknowns and hypotheticals is required, which from a finance lens, may not be prudent. Also, avoidance costs may not be considered – the current solution might require many manual interventions or having to do specific activities inefficiently, which the new solution eliminates. The payback period, which gives a good indication of time the technology will take to pay for itself, doesn't consider change in value of money and is set in stone. It is usually evaluated at the business case stage only.

When considering the levers of supply chain (discussed in Chapter 9), the number of metrics to evaluate can increase exponentially and could possibly be lost when trying to determine financials that you can solely pinpoint to the new solution. Benefits rarely occur from day one of using the new capability, either due to needing to embed the solution or giving time to the users to work with it. Also, importantly, some of the metrics might take time to adjust to new ways of working, such as if a benefit is working capital, a new system on day one is not going to miraculously reduce inventory levels.[2] The new solution is only one part of the jigsaw in how a supply chain will perform, so isolating metrics specific to it will be subjective, which presents challenges. Investments either improve user knowledge, which in turn helps make improved operational decisions[3] or enhances / replaces human activities with automation / robotics.[4]

Efficiency measures

Efficiency measuring usually happens by tracking specific metrics, including lead time, inventory turns and order fulfilment rates. In relation to lead time, consider whether the new solution has led to reduced lead time or helped the company at least maintain the current lead time, therefore reducing failures that can be applicable both upstream and downstream in the supply chain[5] (i.e. supply, own manufacturing and customer).

Inventory turns look at how often inventory is turning from a sold, used and replaced perspective. It can be interesting to understand how well your company is doing against competitors and the industry by benchmarking against them using this metric. By implementing a new technology solution that assists users in managing inventory, you can see whether this is improving (since I imagine in the business case, you will have indicated that it will improve). For example, working with a UK retailer, I was part of a team that introduced a visualization solution that took data from systems and began measuring the performance of identified trouble items. By doing this, it provided users with a focus on improving the turns with recommendations of what to do and targets, leading to tangible improvements in SKU and categories performance.

Measuring inventory days looks at how long inventory is sitting within the business and gives a good indicator of how likely an item will take to sell.[6] Significantly, inventory can tie up the company's working capital and prevent it from having access to cash. If a solution is being deployed here, the expectation would be that inventory days are being measured with a plan to get it within a targeted range. Range is dependent on the type of item that is being sold. For example, you would not want perishable items, such as fruit and vegetables, sitting for too long or they would have to be scrapped, versus a clothing item which might be appropriate for one season or two, versus a spare tyre which could sit for six months plus.

Stockouts are where you are unable to fulfil customer orders due to a lack of stock. There can be a variety of reasons beyond your control to why this happens, from supplier issues to theft. However, this can be helped by implementing a new planning solution that improves forecast accuracy by using more variables to inform demand decisions. This will be more accurate in what you need to fulfil orders, which could be an intangible benefit in relation to the business case.

Order fulfilment looks at how well the business delivers what the customer ordered. This can be expanded to be referred to as on time in full (OTIF), which is a measure that aligns with overall customer satisfaction and if in a retail context, on-shelf availability (OSA) measures whether it was available for consumers to purchase. Most technology solutions will be aligned with improving this metric.

Customer satisfaction

Customer satisfaction, which we have just touched on, is a key metric for any supply chain initiative[7] and measuring its success by understanding how well the supply chain is operating in meeting on-time deliveries is a good indicator of overall performance. It may be more costly to improve customer satisfaction because to meet the on-time element might require the use of different forms of transport to accelerate delivery. However, the new solution should be doing an activity that is linked to helping improve this metric. Order accuracy is another element that the solution should be improving the accuracy of. For example, in a warehouse, a robotic arm can be used to prepare an order for shipment much faster than a human would be able to.

If the company can improve the two elements of on-time and order accuracy with the introduction of a new solution, then customer complaints linked to delivery should reduce. Any further complaints may steer the company to where future solutions may be required (such as in quality) if the customer is still rejecting orders.

Warehouse productivity and throughput

Warehouse productivity can be improved via an appropriate technology solution such as robotics and other forms of automation.[8] One metric that could be of interest to improve and drive benefits could be warehouse throughput. This looks at how quickly the company can move items through the warehouse, from receiving them, to storage, to pick and pack, then shipping. It can even break down each area to see where bottlenecks are and where the use of automation could speed up specific activities. This type of metric is important because it gives a feel for how well the operation is doing and how it could be improved further. The use of automation rather than staff doing specific tasks is justifiable as it leads to many benefits. For example, putaway when done manually, requires several movements by people and the use of forklifts. This could be done more quickly and safely with automation that uses conveyors from the receiving area and then automatically moves products into the storage system – this can also be an automated storage and retrieval system.

By having different solutions, the warehouse capacity could be improved, which means being able to handle more items. This could lead to potential benefits such as supplier collaboration and reducing stockouts.

Transport orchestration

Transport orchestration metrics that you could be interested in improving might include using systems to automate the tendering process for route and carrier selection. This leads to less manual activities being done by staff and potentially identifying the cheapest freight costs for the route desired.[9] By providing more visibility earlier on routes, you could combine deliveries and therefore reduce fuel costs. Also, through having such a system, you might be able to reduce the time and cost of users having to engage with carriers and plan routes, therefore deploying users on other activities. A further benefit could be implementing a network solution where you are able to control deliveries and routes to reduce overall cost to serve.

Technology capabilities

Now that we have hovered over several metrics that could be applicable across the supply chain, let's deep dive into some of the technology capabilities that are typically considered for investment and the measurable benefits they can bring.

Enterprise resourcing planning (ERP)

Enterprise resourcing planning (ERP) systems are thought of as the backbone for any company,[10] which helps companies to integrate and manage many functional activities (e.g. finance, HR, customer service, supply chain). They host the main dataset which can be used by each function and works across functions. They were traditionally the system of choice to run all parts of the supply chain. As the complexity of running global supply chains evolved, their use has become consistent as providing a source of truth at the transactional level.[11] Many ERP vendors have begun to look at the supply chain as modules that can be implemented as standalone or can connect with other vendor solutions, such as in the planning domain where many alternatives now exist.[12]

Benefits of implementing an ERP within the supply chain can be based on how much supply chain capability you have within it:

- **Visibility** – provides tracking and monitoring of inventory, shipments and orders.
- **Transparency** – holds transactional data showing what products are in the supply chain at various points of their transformation, from raw materials through transformation to finished goods, which enables all relevant users to access information and make decisions.
- **Optimize inventory** – automates inventory tracking, allowing users to optimize inventory levels, reduce stockouts / overstocks.
- **Forecasting** – provides forecasting and demand planning capabilities so companies can anticipate and meet demand.
- **Order processing** – automates processing of orders, allowing companies to quickly and accurately process orders and reduce errors.
- **Order fulfilment** – provides near real time tracking and monitoring of orders, enabling companies to improve fulfilment rates and minimize shipping errors.
- **Supplier management** – provides supplier management capability for users to manage supplier transactions, relationships and improve communications.
- **Collaborative planning** – enables planning and forecasting sharing with the supply base so the company can better react and meet demand.
- **Shipping and receiving** – automates shipping and receiving processes to reduce errors and improve efficiency.
- **Routing and scheduling** – routing and scheduling capabilities to optimize delivery times and transportation costs.
- **Automating processes** – many supply chain processes can be automated, reducing manual errors and improve efficiency.
- **Reporting and analytics** – provides reporting and analytics capabilities so the users can understand the operations and identify areas for improvement.
- **Tracking and monitoring** – tracking and monitoring of orders and shipments to assist users resulting in reducing customer uncertainty and complaints.

- **Communication with customers** – have accurate and timely information about orders and shipments.

All of this sounds great; however, based on experience, how much of this is done by user intervention versus the system will vary drastically between companies and ERPs, depending on how they are set up. ERPs aren't a catch-all for supply chains.[13]

The key metrics to evaluate when considering implementing an ERP system (or modernize the current version) for supply chains may include those shown in Figure 10.1.

Figure 10.1 ERP key metrics

- Fill rate
- On time in full (OTIF)
- Freight and transport costs
- Working capital
- Inventory turns
- Days in inventory
- Days sales outstanding
- Days purchasing outstanding
- Costs of ERP legacy systems (hardware and software)
- Costs of ERP legacy systems support

Planning system

Over the last 10 years, a variety of planning technology solutions have come to the market to address the gap in capability seen with the legacy systems, which were usually part of an ERP suite. While many of them refer to themselves as providing an 'advanced planning system', I think the word 'advanced' is misleading. 'Advanced' possibly in that they have more capabilities built in, but these capabilities were executed on spreadsheets manually before, hence my reason for questioning the suggestion of a dramatic change (i.e. codifying what we were already doing seems more appropriate).

Planning systems in the market today can interact with the ERP (which typically holds the master data), ingesting data to help analyse, optimize and inform planning decisions. The new systems have formalized the use of statistical models and embedded these so the user can choose between them when looking at forecasting (many of these models – it should be noted – were developed in the 1960s),[14] providing different approaches to

Figure 10.2 Planning system key metrics examples

- Forecast accuracy
- Exceptional orders
- Adherence to plan
- Inventory turns
- Inventory obsolete rate
- Inventory carrying costs
- Order fulfilment rate
- Lead time
- Service level agreement compliance

that process. These can then be used in conjunction with traditional approaches to forecasting, such as historical sales, seasonal sales and lost sales, for example. The additional elements, such as bringing in unstructured data[15] (e.g. weather, economic data, local events), are helping to aid decision makers with more informative data and being able to run scenarios quickly should lead to more informed, accelerated decision-making.

Benefits of implementing a planning system from the perspective of supply chain include:

- **Improved forecast accuracy** – by using more sources of data with near real time capability, advanced analytics and mathematics (potentially machine learning algorithms in addition to statistical maths), the ability to have a more informed forecast, which is more accurate to customer demand with reduced human errors, is higher.
- **Move to automation** – elements of the planning processes across the various types of planning (demand, supply, production etc) can be more efficient, reducing time to form a view and reducing human errors.
- **'What if'** – simulating and modelling different scenarios to optimize the supply chain, reduce costs and identify potential risks and develop mitigation strategies to resolve.
- **Collaboration** – near real time visibility of planning enables users to respond to decisions with accurate data and engage other parties (e.g. suppliers and customers) to improve the overall supply chain performance and reduce errors.
- **Inventory understood** – tracking of inventory in near real time, automating balancing of inventory across the supply chain from levels, replenish and allocate perspectives, leading to reducing stockouts and overstocking.

- **Reporting and analytics** – provide capabilities to better understand in near real time what is happening across the supply chain from the planning perspective and complete S&OP activities quicker.

The important metrics for implementing a planning system for supply chain include those shown in Figure 10.2.

Procurement systems

Procurement systems have typically been bolted on within ERPs. However, procurement systems were one of the first systems to move beyond the ERP with the introduction of a variety of alternative solutions.[16] Demand for this additional capability came from the time-consuming activities buyers had to do, using spreadsheets and other applications to get partial visibility and control of their categories, as well as keeping suppliers' information relevant and updated. Nowadays it is typical to see multiple procurement applications working in tandem, standalone from the ERP, in executing procurement activities.

Benefits of implementing a procurement system from the perspective of supply chain include:

- **Automated processes** – ability to automate many procurement processes, reducing the time and effort of users.
- **Workflow simplification** – streamline workflows (i.e. approval flows) so multiple users can engage in a workflow and know when to review and approve, reducing downtime and effort required to manage procurement activities.
- **Supplier management** – ability to have relevant documentation linked to suppliers available for users to engage with, understand spend and products they supply, enabling leverage on pricing and terms.
- **Control spend** – reduce maverick spend, number of suppliers and control who the company does business with, providing leverage via spend consolidation.
- **Compliance** – help companies comply with regulatory requirements on transparency and accountability and have one source for contracts and managing them in a standard way.
- **Improved decision-making** – provides users with near real time data, enabling them to make quicker, informed decisions relating to supply chain activities (e.g. procurement users can see beyond level 1 information on spend, drill down and know where it is being used, inventory positions).

- **Analytics capability** – provide flexible analytics that can help users look at supply from a variety of perspectives to support decision-making.
- **Informed supplier relationships** – improve communication and build stronger, informed relationships.
- **Standardize performance** – enhance performance measurement and metrics with suppliers to help evaluation and selection.
- **Resilience** – assessing health and mitigating risks within the supply base and the ability to respond to disruptions quickly.
- **Visibility** – stakeholders have visibility of procurement activities and the ability to quickly assess any situations in relation to supply.
- **Assurance** – having all procurement activities executed in the system provides a comprehensive audit trail and control over spend.

The important metrics to implement a procurement system for supply chain are shown in Figure 10.3.

Figure 10.3 Procurement system key metrics examples

- Spend on direct and indirect materials
- Spend under management
- Number of suppliers
- Days payables outstanding
- Supplier performance
- Compliance to contract
- Procurement labour productivity

Manufacturing execution system (MES)

A manufacturing execution system (MES) helps companies that manufacture products to manage and optimize their production processes.[17] Within the scope of the system can be planning, scheduling and quality management through to managing the execution of production, providing visibility of the operations in near real time.

Benefits of implementing a MES from the perspective of supply chain include:

- **Forecasting and planning** – provides near real time data on production matters (such as capacity, material availability, schedules), enabling improved forecasting and planning of production.

- **Here and now** – users can see how well production is going in near real time, understanding and planning for changeovers and identifying any issues with production.
- **Optimization** – can optimize schedules for production, improving lead times, reducing waste and improving efficiency of operations.
- **Analytics** – helps users understand what problems occur with production and can review to solve issues and improve efficiency of sequencing of production.
- **Inventory** – supports users in understanding when inventory is consumed and/or transformed into work-in-progress and finished goods in near real time, plus showing waste / scrap or reusable inventory.
- **Visibility** – assists users in executing production and what was produced on a run / shift, improving batch output and identifying issues with specific raw materials / packaging inputs in here and now to inform decisions.
- **Communication** – facilitates communication between production and other subfunctions of the supply chain, ensuring alignment and information shared to trigger other parts of supply chain (both up / downstream) to make decisions.
- **Quality** – provides monitoring in near real time of quality control, enabling speedy identification and resolution of issues.
- **Compliance** – tracking and recording production data results in any recalls or issues are able to be quickly identified and actioned.
- **Efficiency** – optimizes the production process to drive overall operational effectiveness.
- **Variability** – helps minimize production variability by having near real time visibility of operations and the ability to enable corrective action.

The important metrics for implementing an MES for supply chain include those in Figure 10.4.

Figure 10.4 MES key metrics examples

- Overall equipment effectiveness
- Production output
- Available capacity utilization
- Plant maintenance costs
- Energy consumption
- Manufacturing labour productivity
- Costs of scrap and rework
- Service level agreement compliance

Inventory management systems

Inventory management systems provide companies with the capability to help manage inventory levels, track inventory movements and optimize inventory across the supply chain. When used well, they can help a company control how it uses inventory within parameters and focus the user on exceptions that the system has flagged due to not aligning with what is required for production, for an example. This might be done for strategic reasons, such as ramping up inventory levels before a marketing promotion or forcing down inventory to necessitate an increase in price linked to demand.

Tracking of inventory when done through transactional systems (e.g. ERP) might be correct within a 12-hour window but might not assist the user who needs to see inventory tracking in near real time, such as in stores when offering customers a variety of opportunities to purchase through a website (e.g. click and collect).[18] Furthermore, finding a balance of where to put inventory is a challenge for retailers whose traditional approach was to have inventory in all stores. It takes time to understand local nuances and then stock stores with products that sell well in those areas. This returns to the power of being able to segment and look at inventory in different ways (e.g. analyse and treat inventory differently rather than via standard ERP classifications of minimum / maximum levels); the ability to do that well requires dedicated capabilities, which can be found in inventory management systems.

Benefits of implementing an inventory management system from the perspective of supply chain include:

- **Inventory accuracy** – provides near real time tracking and monitoring of inventory levels, helping companies make informed proactive decisions and reduce errors with accurate information with little human intervention.
- **Automated data** – can automate data collection and keep this up to date, which previously would require many stock takes, reducing errors from manual entries.
- **Reduce stockouts / overstocking** – using near real time inventory data means knowing the impact of changes in demand and responding quickly, such as triggering dynamic pricing to clear items.
- **Optimization** – have a clear view on inventory levels and recommending alternatives to keep inventory turning, such as changing the lead time of fulfilment, reacting to demand variability and supplier reliability.
- **Minimize lost sales** – ability to react to stockouts and move inventory to sales positions quickly through extraordinary shipments.

- **Minimize costs of overstocking** – identifies where inventory is not turning and enables users to then do something different with inventory, such as moving it through the network or dynamic pricing.
- **Simplify inventory processes** – reduce the time and effort to manage inventory and make inventory processes standardized.
- **Collaboration** – accelerate time to decision between various parties on what to do with inventory issues, including using chat and workflows to automate engagement and resolution.
- **Analytics** – provide analytics capabilities to analyse inventory data, identifying trends and patterns.
- **Order fulfilment** – reduce time and effort to fulfil customer orders by having accurate inventory information.
- **Availability** – resolve inventory issues quickly and have improved availability, leading to a reduction in the likelihood of stockouts, resulting in improved customer satisfaction.

Figure 10.5 shows the important metrics to implement an inventory management system for supply chain.

Figure 10.5 Inventory system key metrics examples

- Working capital
- Cycle stock cost
- Safety stock cost
- Costs of scrap and rework
- Inventory carrying costs
- Inventory turns
- Days in inventory
- Days purchasing outstanding
- Spares and maintenance, repairs and operations (MRO) inventory
- Lost sales cost
- Labour productivity associated with inventory management

Order management system (OMS)

The order management system (OMS) has been included in this chapter because while it has been most prevalent in retail, it is beginning to be considered in servicing sectors beyond. The solution enables companies to manage and fulfil customer orders across multiple channels (e.g. online, click and collect etc) and fulfilment locations (e.g. in-store, locker, distributor etc).[19] In a supply chain context, it ensures orders are processed efficiently, accurately and in a timely manner. Also, it can store all your customer details and can be used to provide alternative customer journeys based on the perceived loyalty the company bestows on the customer (based on spend, for example).

The benefits of implementing an order management system (OMS) from the perspective of supply chain include:

- **Automating orders** – order processing and fulfilment as well as preferred locations are automated, leading to a reduction in errors and an increase in speed to execute orders.
- **Customer experience** – provides the customer with accurate and timely information in relation to their order and the progression through the physical supply chain to them.
- **Visibility** – shows where products are to all relevant parties and where bottlenecks or user intervention is required.
- **Reduces costs** – assists the company in reducing stockouts, overstocking and shipping errors with the automation doing most of the execution of activities.
- **Supports scalability** – as the company's business grows and expands, it can handle increases in order volumes and complexity.

The important metrics for implementing an order management system for supply chain are outlined in Figure 10.6.

Figure 10.6 OMS key metrics examples

- Customer satisfaction rate
- Order fulfilment rate
- Order backorder rate
- Order processing time
- Order cycle time
- Inventory accuracy rate
- Inventory discrepancy rate
- Inventory turnover
- On time in full (shipping and delivery)
- Return rate
- Order cancellation rate
- Order change rate

Warehouse management system (WMS)

Warehouse management systems (WMS) help companies manage and optimize their warehouse operations, from inbound activities (pre-receiving, receiving and putaway and returns) to outbound activities (order management, picking and packing, and loading and packing), including managing inventory, handling trade and reporting.[20] They are also the conduit for implementing automation and robotics in warehouse operations whereby control systems of these solutions integrate with the WMS and then the WMS integrates with other company systems such as ERP, OMS and TMS, depending on the activities the company does.

Benefits of implementing a WMS from the perspective of supply chain include:

- **Accuracy of inventory** – provides near real time tracking and monitoring of inventory in the warehouse operations, providing users with accurate information to inform decisions.
- **Automated data** – movements through the warehouse operation are heavily automated, eliminating manual errors.
- **Simplify processes** – streamline processes in the warehouse, reducing time and effort to manage inventory.
- **Productivity** – automating tasks and optimizing workflows reduces the need for manual labour and allocates resources to utilize labour.
- **Fulfilment of orders** – speed of fulfilling orders is vastly improved, reducing time and effort to create pick lists for humans and machines to orchestrate orders.
- **Availability** – accelerates product availability to other systems, therefore reducing the likelihood of stockouts.
- **Near real time data** – enables users to make decisions on priorities and reduces the risk of inventory being unavailable.
- **Analytics** – supports users with data to identify trends and patterns to improve operations in the warehouse and report out performance in near real time.
- **Compliance** – helps companies comply with regulatory requirements by having inventory tracked and reported at all stages of warehouse operations.
- **Inventory costs** – reduces costs associated with lost sales and having to action emergency shipments.

- **Collaboration** – visibility of inventory enables other functions and suppliers to respond quickly to changes in customer demand and requirements.

Figure 10.7 outlines the important metrics for implementing a WMS for supply chain.

Figure 10.7 WMS key metrics examples

- Available to promise
- Throughput
- Order cycle time
- On time in full
- Capacity utilization
- Inventory carrying costs
- Inventory return rate
- Backorder rate
- Order errors
- Correct documentation
- Cycle count accuracy
- Warehouse labour productivity:
    - o Lines received and put away per hour
    - o Lines picked and shipped per person per hour
    - o Orders picked and shipped per person per hour

Transportation management system (TMS)

A transportation management system (TMS) assists companies in organizing and efficiently managing their transport operations.[21] The scope of the solution can vary but it should include load planning (optimizing loads to maximize capacity and reduce empty miles), optimize routing (dependent on whether the company owns delivery or outsources it to a logistics provider), be the home of carrier management (managing carriers with contracts, rates and performance) and shipment tracking.

Benefits of implementing a TMS from the perspective of supply chain include:

- **Processes and workflows** – automate processes and simplify workflows to reduce time and effort to manage transportation.
- **Costs** – reduce costs to service customers through route optimization and selection of carriers to execute activities.

- **Visibility and tracking** – provides users with near real time tracking to monitor shipments and inventory from leaving their operations to customers as well as visibility to all interested parties.
- **Data and analytics** – near real time data allows users to proactively make decisions based on situations that unfold, such as breakdown or carrier failure.
- **Safety and security** – help companies manage relevant compliance in relation to these topics.
- **Contract management** – enables users to manage contracts and monitor the performance of carriers consistently.
- **Resilience** – helps companies assess risk with transport, respond to disruptions and potential carrier financial issues.
- **Customer** – improve reliability in servicing customers with knowing what is going on with shipments as well as increasing the speed of execution of tasks to action the movement of products.
- **Communication** – enables all relevant parties to understand what is going on with shipments from leaving the company's operations.

Figure 10.8 shows the important metrics for implementing a TMS for supply chain.

Figure 10.8 TMS key metrics examples

- On time in full
- Late delivery rate
- Freight costs
- Cost per mile
- Trailer utilization rate
- Mode of transport
- Transport labour productivity
- Service level agreement compliance (customer)
- Carrier delivery rate
- Carrier cost per mile
- Load tender acceptance rate

Visibility system

I am cautious to use the phrase 'control tower' (discussed in Chapter 4) to describe a visibility layer[22] (i.e. the ability to see through the connecting of systems and data) due to negative connotations which have come with this capability whereby it promised so much but has not really made the step change that it was expected to. The 'visibility layer' is a relatively new concept that has not found a standard format. However, it is worth discussing since it is becoming the next iteration of control tower and it includes capabilities such as natural language processing (NLP), providing users the ability to ask the system questions and the ability to bring other systems' data together to orchestrate supply chain operations. We will discuss this in much more detail in further chapters when looking to the future.

The benefits of implementing a visibility system from the perspective of supply chain include:

- **Visibility** – providing near real time data into what is going on across the supply chain to improve decision-making.
- **Scenario management** – ability to run scenarios and understand impact to improve decision-making, considering other factors beyond physical movements such as cost and time.
- **Collaboration** – encourages collaboration across global supply chains and outside parties (i.e. multi-party collaboration), working together to optimize operations and improve performance.
- **Communication** – provides users with the ability to engage on a system with others, pinpoint activities and actions via war rooms to solve specific issues and auditable outputs to improve decision-making.
- **Automation** – processes can be automated to reduce the time and effort required to gather information and act on it.
- **Workflows** – brings together workflows that are typically siloed through having visibility to see across supply chain and zoom in on specific areas of concern to solve.
- **Data** – near real time data enables users to quickly engage with a situation.
- **Analytics** – the ability to run reports and see what is going on across the supply chain, identifying trends and patterns.
- **Resilience** – spotting risks becomes quicker and users can see how it is affecting supply chain performance and react.
- **Agility** – ability to respond in near real time and act decisively with facts.
- **Flexibility** – enables the company to adapt to change and put in place actions when issues crop up.
- **Customer experience** – enhances reputation with customers, since the company is able to share information with them and help them be part of decision-making if something adverse occurs in the supply chain.

- **Communication** – encourages engagement across functions and with third parties, observing and acting based on what is always going on within the supply chain.

The important metrics for implementing a visibility system for supply chain are shown in Figure 10.9.

What becomes apparent when looking at the benefits and metrics that are important to measure for all these systems is that there is much crossover and many different factors at play, which can determine the success of a technology investment in supply chain. One should be cautious to suggest that the technology alone drove the benefits.

Figure 10.9 Visibility key metrics examples

- Return on capital employed (ROCE)
- Cost to serve
- Working capital
- Forecast accuracy
- On time in full
- Transport exceptions

Practices for measuring success

Measuring success, as shown already in this chapter, can go in a variety of directions and can sometimes be hidden when trying to draw a line from investment to benefits for a company. Sometimes the new solution is required for reasons beyond financial benefits or can be difficult to quantify, i.e. because of changes in laws or vendors' sunsetting support for legacy systems.

At the outset of developing the case for the new solution, it is imperative to state the goals and objectives for why this technology investment is required and also what metrics will change. It can be difficult to sometimes explicitly state these; however, engaging with benchmark data and vendors, and reviewing current key performance indicators (KPIs) should help with the creation of an initial list, which can be refined.

Defining KPIs should include financial metrics, including ROI, TCO and payback period. However, from there, what the solution can impact should be assessed. For example, while implementing a new procurement solution will drive improvements in engagement with suppliers (that will then lead to reduced costs and a more responsive supply base), linking that to order fulfilment might be difficult due to the many variables that can affect that metric through other parts of your supply chain.

Data and the ability to track and analyse it (as it affects metrics of success) should be used across the various capabilities mentioned previously – in relation to efficiency, customer satisfaction and inventory. Visualization (i.e. the ability to see data in a user-friendly way, such as dashboards) is a critical element of all these different capabilities because tracking the supply chain and how it is performing now compared to before should be bolstered by linking technology investments to improvements in these metrics.

Ongoing monitoring, evaluating and adjusting the technology where appropriate should be carried out through the life of the business case with periodic reviews, ensuring that it is delivering against the desired milestones and outcome. This will aid the supply chain leadership in being able to demonstrate value and support future investments in technology, such as the enhancement of the solution with further capabilities.

This has been alluded to already, but data that measures the success of the investment can be tricky to locate due to it being part of the orchestration of the supply chain. How to isolate it and hold the mirror up to the investment can be a subjective distraction. This can also be the case with defining KPIs, which can strictly judge whether it is delivering the perceived benefits. Also, any ongoing measuring will require resource time and effort, which the company may not have available.

EXPERT INSIGHTS

Other experts I spoke to provided some concluding views on measuring success.

The importance of evidence-based decision-making within the transformation process, the need for continuous evaluation of initiatives and the role of leadership in fostering a culture of innovation and adaptability is crucial.

Tracking and maximizing the value of the supply chain requires ownership. Staff need to champion it and thoroughly examine and take ownership of their responsibilities.

Also, it's important not to assume that a new supply chain technology capability will solve all business issues; to do so is flawed, as it often perpetuates the same problems and cultural issues within the new system. For example, data cannot be created without human intervention and the belief that a new data and analytics system will meet all data requirements is misguided. Without creating the data, there is no data to analyse. Therefore, it's crucial to ensure that the foundations or a clear action-based strategy of change are in place before implementing a new system in the supply chain; this will enable the organization to utilize and deliver the intended benefits effectively.

Successful technology implementation prioritizes getting the business back up and running with minimal disruption instead of hitting precise financial targets, especially valuing operational continuity and stakeholder satisfaction over financial metrics. A reality check is sometimes required: many businesses are far less advanced than people assume and practical experience often trumps theoretical approaches in supply chain transformation.

Lastly, don't over-promise on the speed of delivery of the benefits. It will take at least 12 months for most new supply chain systems to ramp up to 'full run rate' of savings as it takes time for people to understand and deliver on the new processes.

Conclusion

The ability to measure how well a supply chain technology investment has done is crucial to maintaining confidence in the supply chain and its ongoing performance. Technology investments in the supply chain involve a significant amount of money and time; therefore their ability to help improve the supply chain requires ongoing justification, but if successful, can support future technology investments. There are many benefits that aren't quantifiable, such as companies that use modern technology in their supply chains attracting a higher calibre of recruits.[23] Implementing one solution usually helps to drive improvements in one or two areas, however this can also highlight where other issues are that the technology could improve. The difficult part is developing the strategy and then having the support to implement appropriate capabilities, keeping in mind the overarching mission. By having clear metrics, which can be attributable to the solution and tracking them as an ongoing activity following best practices, companies can overcome challenges and understand the success of their investment.

What can challenge gaining the potential benefits is the context of the company and at what stage it is in evolving its supply chain technology stack. Moving from legacy systems to modern systems should drive a different benefits profile than incremental changes or bringing in disruptive solutions. Remember that calculations should be based on what you can measure rather than a range of savings that you can't point to.

This chapter has looked at a variety of solutions in the supply chain sphere; however, it is not meant to be exhaustive and as we will look at in Chapter 13, several disruptors are appearing that could change which technology capabilities may be best suited to providing companies with competitive advantage in their supply chains in the future.

Endnotes

1 Williamson, C (2025) Podcast Episode 904 – Rory Sutherland – The secret weapons of marketing psychology, *Modern Wisdom*, https://player.fm/series/modern-wisdom/ep-904-rory-sutherland-the-secret-weapons-of-marketing-psychology (archived at https://perma.cc/3Z56-TZMJ)

2 Foster, T A (2023) Making the ROI case for IT investment, Supply Chain Brain, www.supplychainbrain.com/articles/5132-making-the-roi-case-for-it-investment (archived at https://perma.cc/RY56-QAYX)

3 Blankley, A (2008) A conceptual model for evaluating the financial impact of supply chain management technology investments, *The International Journal of Logistics Management*, ISSN: 0957-4093, www.emerald.com/insight/content/doi/10.1108/09574090810895942/full/html (archived at https://perma.cc/F4EX-AXUJ)

4 Patel, V (2020) Automation, robots and A.I: The rise of the supply chain machines, Supply Chain Digital, https://supplychaindigital.com/digital-supply-chain/automation-robots-and-i-rise-supply-chain-machines (archived at https://perma.cc/XK8W-MSAB)

5 Jenkins, A (2022) Supply chain dfficiency: Definitions, metrics and steps to improve, Oracle Netsuite, www.netsuite.com/portal/resource/articles/inventory-management/supply-chain-efficiency.shtml (archived at https://perma.cc/J8QV-PCRJ)

6 Blanchard, D (2021) *Supply Chain Management Best Practices*, 3rd edition, Wiley

7 Agyabeng-Mensah, Y, Ahenkorah, E N K and Osei, E (2019) Impact of logistics information technology on organisational performance: Mediating role of supply chain integration and customer satisfaction, *Journal of Supply Chain Management Systems*, www.academia.edu/46130449/Impact_of_Logistics_Information_Technology_on_Organisational_Performance_Mediating_Role_of_Supply_Chain_Integration_and_Customer_Satisfaction (archived at https://perma.cc/3BX3-V6P7)

8 Pandey, D, Pandey, B K Kumar Kanike, U, George, A S and Kaur, P (2024) *Utilization of AI Technology in Supply Chain Management*, IGI Global Scientific Publishing

9 Thomas, J, Tandon, R, Vedi, K V and Gupta, S (2024) Assessing effects on automation in enhancing supply chain performance within the logistics sector, International Conference on Artificial Intelligence and Emerging Technology (Global AI Summit), https://ieeexplore.ieee.org/abstract/document/10947853 (archived at https://perma.cc/G6GW-RKCM)

10 Meiryani, Fernando, B, Hendratno, S P, Kriswanto and Wifasari, S (2021) Enterprise resource planning systems: The business backbone, Association for Computing Machinery, https://dl.acm.org/doi/abs/10.1145/3466029.3466049 (archived at https://perma.cc/8HGP-LWES)

11 Dua, P (2024) How master data management can enhance your ERP solution, Stibo Systems, www.stibosystems.com/blog/how-master-data-management-can-enhance-your-erp-solution (archived at https://perma.cc/A4TH-7Y7A)

12 Gartner (2024) Gartner magic quadrant for supply chain planning solutions, Gartner, www.gartner.com/en/documents/5374263 (archived at https://perma.cc/WL4J-4CC4)

13 Grabler, T (2021) Your ERP system isn't the best tool for supply chain management – here's why, TOPO, https://topo.cc/your-erp-system-isnt-the-best-tool-for-supply-chain-management-heres-why (archived at https://perma.cc/8DWS-8GTW)

14 Brown, R G (1963) *Smoothing, Forecasting and Prediction of Discrete Time Series*, Prentice-Hall International

15 Nguyen, H D, Tran, K P, Thomassey, S and Hamad, M (2021) Forecasting and anomaly detection approaches using LSTM and LSTM autoencoder techniques with the applications in supply chain management, International *Journal of Information Management*, 57, ISSN 0268-4012, www.sciencedirect.com/science/article/abs/pii/S026840122031481X (archived at https://perma.cc/A3HP-H2P2)

16 Joshi, N (2025) How is the e-procurement system different from ERPs? Procure Insights.AI, https://procureinsights.ai/e-procurement-vs-erp (archived at https://perma.cc/8G7E-WVW9)

17 Shojaeinasab, A, Charter, T, Jalayer, M, Khadivi, M, Ogunfowora, O, Raiyani, N, Yaghoubi, M, and Najjaran, H (2022) Intelligent manufacturing execution systems: A systematic review, *Journal of Manufacturing Systems*, 62, ISSN 0278-6125, www.sciencedirect.com/science/article/abs/pii/S0278612522000048 (archived at https://perma.cc/3ZHU-H3NS)

18 Thomas, C (2024) Optimizing your inventory management: Why your ERP system may not be enough, ASC Software, https://ascsoftware.com/blog/erp-in-inventory-management (archived at https://perma.cc/827U-RGUS)

19 Pagel, G (2023) What is an order management system (OMS)? Planet, www.weareplanet.com/blog/why-order-management-systems-important (archived at https://perma.cc/2P6V-UL7N)

20 Khan, G M, Huda, U N and Khaleeq Uz, Z (2022) Smart warehouse management system: Architecture, real-time implementation and prototype design, Special issue: Intelligent Factory 4.0: Advanced Production and Automation Systems, www.mdpi.com/2075-1702/10/2/150 (archived at https://perma.cc/VBC7-YCGG)

21 Drljača, M and Sesar, V (2023) Supply chain transportation management, *Transportation Research Procedia*, 74, www.sciencedirect.com/science/article/pii/S2352146523004490 (archived at https://perma.cc/Q4FR-S2FJ)

22 Adams, M (2024) What is supply chain visibility & why does it matter? Z2Data Team, www.z2data.com/insights/what-is-supply-chain-visibility-why-does-it-matter (archived at https://perma.cc/WFJ3-QKKA)

23 Gartner (2024) The Gartner Supply Chain Top 25 for 2024, Gartner, www.gartner.com/en/supply-chain/research/supply-chain-top-25 (archived at https://perma.cc/76PL-UB3Y)

11 The role of people in supply chain technology adoption and implementation

This chapter will explore the role of people in supply chain technology adoption and implementation, including training and development strategies and change management.

> **CHAPTER AIMS**
>
> - Understand the importance of people in technology adoption and implementation.
> - Learn why training and development strategies and change management are essential for technology adoption.

'Technology is nothing. What's important is that we have faith in people. That they're basically good and smart, and if you give them tools, they'll do wonderful things with them.'

STEVE JOBS[1]

REAL-WORLD EXAMPLE Lidl

In 2018, it was revealed that Lidl, one of the leading discount grocery retailers had been grappling with the implementation of a new ERP system for just over seven years and with costs of €500 million, it decided to abandon it. Its story drives home the importance of a people-centric approach to implementing new technology solutions.[2] The project, like so many before and after, was about scrapping legacy systems, infusing modern technology that would standardize operations across countries and deliver significant benefits to the company. From a supply chain perspective, the company was expecting to benefit from this new and improved ERP system, with a better control of inventory high on the executives' priority list, as well as the usual themes of improved efficiency, streamlining of operations and a system that could handle the expected growth, which was steadily being achieved.

The implementation went wrong from a people perspective, demonstrating lots of factors as to why this was, which we will cover later in this chapter. Firstly, resistance to changing business processes was evident from the change from the legacy system that based purchasing goods at the actual price (the new system used standardized retail pricing). Employees pushed back against this, trying to force old ways of working on the new system, leading to expensive and time-consuming customizations. Buy-in from employees was also a struggle with limited training provided to get familiar with the new system and many did not see why they were being asked to use new technology (i.e. the vision and the benefits were poorly articulated). Face-offs occurred between business users and the IT department, whereby priorities were not aligned; employees wanted to keep working the same way as they had always done versus the IT department, which wanted standardized processes. This resulted in friction, significant additional cost and time-consuming delays and modifications to the system. The underestimation of change management became apparent throughout the period, with too much focus on the technology. With the assumption that the business users would adapt, leadership failed to manage the company through the change. The key takeaway from this is that technology alone does not solve problems: people working with strong processes, underpinned with solid data and technology do.[3]

As Steve Jobs articulated in the quote at the start of this chapter, as important as technology is, it means very little if your people aren't equipped to use it to do their roles. Where technology stands the test of time is in the adoption and daily use once it has been implemented. How many times have you seen users work off spreadsheets rather than using the systems provided and creatively develop workarounds that add further tasks and time to execute processes? This knowledge then sits with the user, so when that person leaves, the knowledge walks out the door with them.

The implementation and adoption of new technology in supply chains is a complex, challenging process that requires detailed planning, execution and ongoing support. Although it is a crucial component of supply chain management, its success or failure relies heavily on the people involved.

The human factor in technology adoption refers to the social, psychological and cultural aspects of technology adoption. It encompasses the attitudes, behaviours and interactions of individuals and groups within a company, as well as the external stakeholders, such as customers and suppliers. The human factor can either facilitate or hinder the adoption of new technology and it is essential to address the human aspects of technology adoption to ensure successful implementation.[4]

When we think of adoption, there are several human factors that jump to mind. Topics we keep coming back to time and again in this book are resistance to change,[5] caused by a lack of understanding of the technology and the reason for the change, and the fear around a person's job security or a change in responsibilities.[6]

We will explore each of these in more detail during this chapter, as well as looking at why training and development are important,[7] not just for those using the technology daily but for those with a key role in implementation. From there, we will explore how to create strategies to effectively provide training and development (which sometimes can feel like a tick-the-box exercise rather than it being a journey) so that employees can become experts in the technology capability. Then we will reflect on change management and communication, outlining approaches of how to get value from these activities.

Managing a legacy solution and transitioning over to a new solution will require a detailed plan and stakeholders to remain focused until the transition has occurred, and the new solution has been successfully implemented. It is common near the end of the implementation to forget about the importance of remaining operational and fulfilling customer orders through the old system. Therefore, exit criteria must be established for when users begin to migrate over to the new system and this needs to be communicated with relevant parties (e.g. customers, logistic providers) to support the transition. Usually, if communication is strong between parties, inventory front loading in the short term is acceptable to give the company time to settle in with the new system. I have seen countless times where inventory is positioned to let the business ease into a new system with reduced orders for the first few days and then ramped up.

Managing human factors better

To say that human factors that can influence the outcome of a technology implementation and adoption in supply chain are table stakes is an understatement.[8] The initial and ongoing success of technology in supply chains depends on how a company moves from its current state to then identifying the need that a technology capability can provide that will make a difference to the performance of the supply chain. Some companies only make the change based on factors beyond their control, such as vendors sunsetting or due to a wider technology upgrade and it becomes apparent that the supply chain technology cannot integrate with it or the cost of keeping legacy systems maintained is too high. Once the 'need' is identified, it is about rallying supply chain personnel around a future that involves a new technology and being able to articulate the reasons for it. Then, consideration must turn to how are we going to implement and embed this new technology capability so that it delivers the value we expect it to? This is where the initial excitement about the concept quickly goes, as the difficulty of implementation and holding onto the vision is encountered. Change is not easy, however, there are conscious decisions we can make that can help us manage these human factors, which will naturally appear at all stages.

- **Expect resistance.** It is a stereotype to say all people are resistant to change but it is better to go in with that mindset than hoping your leadership through words and actions alone can prevent it from occurring. Adopting a new technology is perceived in many instances as a threat to job security, status or comfort levels. It can manifest in various ways, such as negativity, scepticism and even sabotage as outlined earlier in the book. It can also come from a subconscious place within people where they don't even realize that they are demonstrating negative behaviours.[9]
- **Spend time understanding people's views.** A lack of understanding and/or nervousness about the new supply chain technology can be another element that can prevent people from engaging, from not understanding

the benefits and capabilities of this technology, to a lack of training, poor communication and inconsistent sharing of information. A lack of understanding and nervousness are natural, however, they can be difficult to quantify and know if they have been addressed unless you spend time understanding people's views.[10]

- **Be aware of knowledge gaps.** The adoption of new technology might require people to have specialized technical skills that are not available within the company or take considerable time to develop, which may not be feasible in the time required. This could result in frustration, errors and/or delays throughout the different phases of implementation and make adoption difficult.[11]

Training and development

Essential to supply chain technology adoption is training and development. These activities can help to address many of the human factors that can influence its success or failure. Any project to introduce a new technology solution in the supply chain needs a focused approach of activities to upskill personnel to capture perceived benefits, equipping people to not just use the system but use all the capabilities within it to improve company performance. Most technology solutions should be aiding the user to do their role more productively and provide capabilities that they were unable to do previously, were time-consuming or required them to do activities off systems.[12]

Training and development should never be an afterthought and doing this early in the project helps users get comfortable and adopt the new solution. Get staff involved and enable them to access it. This plays a significant role in building use cases, understanding their pain points and testing and building a solution that will assist them in doing their role better. This should be at the forefront of the project leadership's mind throughout; users need to be empowered. Having this environment set up early can lead to additional innovation since users will be more at ease to provide their views and challenge the design to help improve useability.

Planning the training and development

Having a well-thought-out, clear and concise plan of activities for training and development will provide benefits to the company and the users of the solution. From the outset, training should be building users' technical skills and expertise to effectively utilize the new technology and understand all the capabilities available to them. There are a variety of ways to approach this but having a 'sandbox' environment[13] available will aid users in working with the system early on, getting familiar with how to navigate it and use it to do the activities required of the job. Also, having training pathways for different users will build technical skills in the IT function, who can understand how to configure the system, get comfortable with the logic it uses and how it reaches its outputs. The importance of IT colleagues understanding the business requirements, features, functionalities and integration requirements from an early stage cannot be overstated. From a data management perspective, relevant personnel should learn requirements of the system in relation to data entry and flows, validation and analysis to maintain accurate and useful data.[14]

Turning again to the users, having training focused on them such as 'day in the life'[15] type scenarios must be done to also help them learn how to do the tasks that they require to execute their role. It can address pain points they may have currently, so from an early stage, they get comfortable with the solution. It can lead to building a group of advocates for the solution, once they see the benefits too. As well as providing their requirements, prior or early in 'design', they should be engaged to understand any process changes needed, since existing workflows will most likely need refreshing. These steps aid with embedding change.

It can be a challenge for users to explore the functionalities of the new solution in 'day in the life' scenarios since many will have their 'day jobs'; however, opportunities should be provided to them because of the value this will add throughout the implementation and adoption process. Doing these activities will enhance collaboration and improve communication between all relevant parties, as the 'us' and 'them' feelings,[16] which can reverberate around a company when some chosen individuals are either dedicated fully or partially to the implementation, will be reduced.

One common approach deployed by companies is 'train the trainer',[17] where users are handpicked to become super users and then, in time, train their colleagues. This approach is encouraged because the person training knows what the business users are doing and if chosen well, can encourage the continuity of the technology usage post-implementation.

Keeping external partners abreast of changes is advised too, since it helps to be proactive in communications in case any issues crop up at 'go live'.[18] Also, there may be interactions between the new solution and external solutions, which need rigorous testing during implementation and potential changes to their ways of working with your company (examples of this could include receiving and confirming sales orders or customer orders, and sending instructions to logistics providers), which might require training.

After implementation, there should always be a phase of continuous improvement, which will require additional training plus one-off training when additional features may be added or developed by the vendor to ensure the company stays up to date.

Making training and development effective

The next question is how to make these training approaches effective because anyone who has been through any form of implementation knows that training standards can be inconsistent and it can also be a contentious activity when budgets are squeezed.

Effective training and development necessitate a flexible approach that engages users by understanding how they want to receive training because we all learn differently. There are four predominant learning styles[19] as can be seen in Figure 11.1.

The challenge presented when coordinating training is that people can be dominant in one learning style or have a mixture of the four.[20]

A considered approach to training and development is to engage users with a mixture of approaches to build understanding and then reinforce it. Early on, it might be more suitable to provide online training and e-learning so that all users are upskilled with a general understanding of the solution (i.e. visual, auditory and kinaesthetic). This type of training provides flexibility and convenience for users to complete at a time suitable for them, but it is important to make sure that users' progress is monitored to make sure they are developing understanding at similar times.

Figure 11.1 Four learning styles

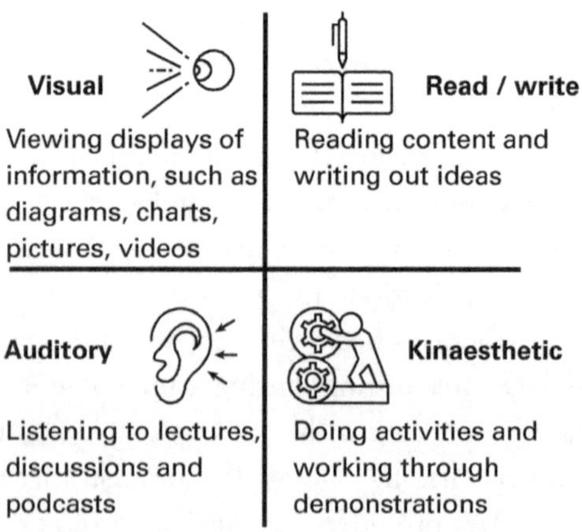

Then it might be appropriate to engage in classroom training (i.e. auditory and read / write) where users can engage in question-and-answer sessions with the technology experts. This also provides early thinking about functionality issues, building a collaborative approach between users where they bounce ideas off each other. It might also make sense to cross-functionally (i.e. not direct users within supply chain, finance, commercial) train users with visual and auditory methods, such as tutorials and demonstrations, who may have limited interaction with the technology or receive information from it, so they understand its benefits and implications for them. For example, these demonstrations could show how to navigate between screens, to see the extent of the technology's capabilities.

There are vendor qualifications available for technical and business users that include a mix of all four styles. These can be helpful early in the project to gain a foundational understanding of the technology and help engage employees, achieving open communication. Achieving qualifications can be morale-building. This is also a good time to deploy specific training for key users and implement 'train the trainer' approaches.

Classroom training (i.e. all four styles) just mentioned will aid the implementation team with the 'design' phase of the technology solution, where users will be discussing requirements and ensuring that these are captured as part of the 'blueprint'. At this stage, it can be helpful to develop user personas, user journeys and wire frames.[21] See Figure 11.2 for examples of these.

From there, it is useful to have regular check-in sessions, providing a 'sandbox' environment with dummy data that users can input to familiarize with the solution (i.e. primarily read and kinaesthetic approaches).

As Figures 11.2 and 11.3 show, creating user personas assists the technology implementation by understanding the importance of each role and what activities the users will do within the solution. User journeys can be used as a checklist against requirements to ensure the solution being developed is 'fit for purpose' early. These help all parties debate whether current workflows need to change or can be improved with the new solution. Also, it builds confidence between the project team and users since it reduces the risk of things getting 'lost in translation', which can happen when developing requirements and subsequent processes.

Involving users in developing scripts for system integration testing (SIT) and user acceptance testing (UAT) will also provide an opportunity for users to engage with the new solution. This will make them feel part of the team to ensure testing has been robust and that the system is performing correctly in the 'test' environment. It is crucial that users are involved heavily at this time, with consideration given to free up their time from day-to-day business to

Figure 11.2 Example of user persona

Country logistics manager:
'Every link in the supply chain matters. My job is to ensure that each one is strong, reliable and future-ready.'

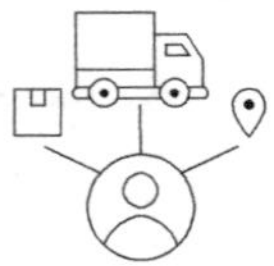

Role:
Responsible for overseeing and optimizing the transportation, distribution and storage of goods within the supply chain. This role entails maintaining cost efficiency and high-quality service standards. They play a pivotal role in driving operational excellence and implementing innovative practices to address evolving industry demands.

Goals:

- Optimize processes
- Enhance co-ordination
- Sustainability initiatives
- Customer satisfaction

Challenges:

- Balancing cost-efficiency with the delivery of high-quality services
- Managing delays caused by external factors

Key motivations:

- Achieving operational excellence and contributing to the success of the organization
- Delivering dependable and efficient logistics solutions to earn client trust and loyalty
- Implementing innovative measures to surpass competitors and solve complex industry challenges
- Making a meaningful impact by incorporating sustainability into logistics operations

Frustrations:

- Customer expectations
- Supply chain disruptions
- Inventory management
- Communication issues
- Regulatory compliance

Expectations of solution:

- Real time tracking: System that provides real time updates on shipments to enhance visibility and responsiveness
- Data analytics: Ability to analyze logistics performance and identify areas for optimization
- Automation: System that minimizes manual tasks
- Integration: System that seamlessly connects with customers and logistics partners
- User-friendly interfaces: Intuitive systems that are easy to navigate, enabling quick adoption by the team
- Sustainability tools: Features that help measure and reduce the environmental impact of logistics activities

Communication style:

- Clear and concise communication for internal and external co-ordination
- Timely updates and structured collaboration through digital platforms for team alignment
- Structured discussions during face-to-face or virtual meetings to address strategic objectives

Figure 11.3 Example of user journey for country logistics manager

Stages	**Planning and forecasting**	**Procurement and sourcing**	**Inventory management**	**Transportation and delivery**	**Customer communication and support**	**Performance analysis and reporting**	**Continuous improvement**
Activities	• Analyze historical data and market trends • Create transportation and capacity plans • Collaborate with sales and procurement to create integrated plan • Assess carrier requirements	• Selecting logistics partners and negotiating contracts • Procuring third-party transportation and warehousing services • Evaluating supplier performance against agreements	• Managing stock levels across multiple warehouses • Ensuring near real time inventory visibility • Reducing carrying costs and stockouts • Optimizing storage locations	• Route planning and fleet management • Ensuring timely and cost-efficient deliveries • Tracking movement to customer • Monitoring third-party logistics (3PL) performance	• Providing shipment visibility to customers • Managing delivery exceptions and customer issues • Aligning logistics performance with customer SLAs • Managing returns	• Measuring delivery times, cost-per-shipment, inventory turnover • Analyzing carrier and warehouse performance • Reporting KPIs to leadership	• Identifying operational inefficiencies • Implementing new technologies or process improvements

Stages	Planning and forecasting	Procurement and sourcing	Inventory management	Transportation and delivery	Customer communication and support	Performance analysis and reporting	Continuous improvement
Pain points	• Inaccurate data or delayed data leading to overstocking / shortages • Demand variability and forecast errors • Difficult and time-consuming to integrate data from different sources to execute from	• Delays in contracting and long onboarding times for new vendors • Difficulty collating data on carriers • Fluctuating rates and hidden costs from suppliers • Limited visibility into supplier capacity and reliability	• Inconsistent inventory data across systems • Difficulty finding inventory • Overstocking due to safety buffers • Manual tracking errors	• Delays in transit • Unexpected disruptions • Difficulty communicating with drivers • Lack of near real-time shipment tracking • Last-mile delivery inefficiencies • Fuel cost volatility and carbon emissions pressure	• Delays in responding to inquiries • Reactive communication with customers • Fragmented systems for support tickets and delivery data • Missed SLAs damaging reputation • Cumbersome returns	• Disparate data sources and lack of automation • Time-consuming manual report generation • Limited insight into root causes of inefficiencies • Lack of KPIs that drive performance	• Resistance to change from teams or partners • Lack of bandwidth for testing and rollout of new tools • Difficulty in measuring long-term financial benefits of improvements

Stages	Planning and forecasting	Procurement and sourcing	Inventory management	Transportation and delivery	Customer communication and support	Performance analysis and reporting	Continuous improvement
Opportunities	• Leverage AI / ML for better planning • Integrate near real time market and supplier data • Cross-functional platform integration for improved visibility	• Use of digital procurement platforms and marketplaces • Dynamic pricing models with transparent benchmarks • Building strategic partnerships instead of transactional ones	• Implement near real time inventory tracking (e.g. RFID, IoT) • Predictive analytics to manage reorder points • Centralized inventory control systems	• Optimize routes using AI / ML • Adopt EVs or sustainable delivery options • Enable live shipment tracking and alerts for stakeholders	• Proactive communication (automated updates, delay alerts) • Customer self-service tracking portals • Centralized 'live' dashboards for support teams	• Use of business intelligence tools and near real time dashboards • Automated anomaly detection • Drill-down analytics for granular decision-making	• Foster a culture of innovation and agility • Use 'pilots' to validate new solutions • Align improvements with sustainability and ESG goals
Feelings	• Cautious • Pressured	• Hopeful • Vigilant	• Stressed • Risk-aware	• High alert • Deadline-driven	• Reactive • Reputation-focused	• Analytical • Accountable	• Curious • Constrained

participate. This is important because after this time, the project will move to cutover (where the organization transitions from one system to the other)[22] and having users' buy-in and training done is needed before this. While many advancements have been made in automating testing, I am of the view that SIT and UAT are value-adding and build confidence and assurance with users, hence it is a worthy activity to spend time on.

Post stabilization of the solution, it is important that users continue to gain training to assist with using all functionality and developing requests for future enhancements. This can be done by providing access to vendor experts and suitable training to keep them updated. Forums with other customers of the vendor solution are helpful, where users can share what they are doing with others and learn from others as well.

So, in summary, the strategies to engage in training and development should be a blend of on-the-job training with classroom, online and e-learning and coaching and mentoring, not forgetting the importance of cross-functional training. Deploying such a mixed approach will help engage users with different learning styles and ensure that the company is prepared for the changeover from an old to new system or the introduction of a new system. To ensure the best outcome, provide appropriate training materials that aid users and don't overwhelm them. Training requirements should be discussed before materials are made available. Also, these become foundational materials to assist new employees when they are onboarded. This is where having key users selected early can help to distil materials and be part of the transformation. Having materials readily available also combats any possible attrition (i.e. people leaving) that naturally occurs within organizations.

Change management and communication

When it comes to transformation involving technology, the ability of companies to get this wrong so often suggests that lessons are not being learned.

Change management starts from the moment the idea of a new technology solution for supply chain becomes something written on paper. It carries on through the development of a 'case for change', throughout the journey and beyond the transformation. When beginning the business case, it is key to have a clear and compelling reason for the change; consider and document what this entails for the company and its relevant parties both upstream and downstream of it. Being able to communicate the benefits and value of the new technology is required and should not just be some bullets on a page, as it is showcasing its importance to the employees. Time spent here will improve traction within the company to get support and begin to move the company quickly out of insecurity – the resistance to change will be present, even if not visible initially.

During the project, frequent and consistent messaging should be the mode of operation. Also, the messaging should not just show the optimistic side, which many projects also hide behind. To think that there won't be bumps along the project path is foolish. Although experience and external support can provide rigour, things can go wrong and it shows humility to explain these and resolve them openly.

Early on, understanding the stakeholder landscape by analysing those who are supporters, neutrals and detractors will help focus the project team on the different types of engagement these parties require and consistent effort should be deployed to engage them. I find that having conversations with them and understanding their needs and concerns helps to build allies in time. Keeping these analyses live and updated is important because sometimes these can feel like one-off exercises (it is also important to remember that people come and go during the project). Also, having these forums allows for discussions about finances and keeping track of progress, so that surprises are reduced and confidence is built in the team. An activity that goes hand in hand with this is communications planning and executing this through the life of the project consistently, discussing different levels of communication, understanding who is best to engage with different stakeholder groups and being transparent with them. One trend that I sadly do not see now as often is the celebration of milestones and building the team spirit through informal get-togethers. This is probably a fallout of hybrid and remote ways of working; however, building a culture of inclusiveness and being part of a mission should not be lost because of this change. Bringing the project team and business users together each quarter and making them feel special can encourage the company to maintain efficiency.

Executing communications simply and effectively early on and being consistent with the approach throughout the project life cycle is critical. Showing the solution and the changes and benefits it will bring need to be continuously repeated and visible in workplaces. Visually showing progress via an overarching plan clearly demonstrates where future users need to be engaged to help. I was involved in a project to implement procurement technology and the project leadership team provided communications at specific milestones. These milestones were celebrated as a wider team, with project merchandise provided as a thank you to all involved. It is also appropriate to update business partners (e.g. customers, suppliers, logistics providers) even if they are not directly affected by the new solution because they can help alleviate potential go-live issues if they are engaged and understand what is happening rather than finding out if an issue occurs last minute.

No project of this kind goes to plan fully, so monitoring and evaluating progress is a daily (if not hourly) activity. Through doing this, challenges will appear much sooner and can be addressed in near real time. The final consideration, which should be conducted post-project, is measuring its success – this can be quite linear (i.e. did we see the tangible benefits we expected?). Reviewing the business case will provide some of the answers; however, its impact can be diluted over time, especially if there has been a change in the C-suite. KPIs and operational performance will provide in-the-moment feedback and over time, are strong indicators of success of the project. The adoption by the users will always be the best indicator of success.

REAL-WORLD EXAMPLE IBM

Within IBM's infrastructure business, its culture has a willingness to try and potentially fail with new technology, with strong leadership support to encourage and support transformation initiatives, empowering its people. On its journey, the challenge is acknowledging the balance of speed versus risk with data accuracy at the centre of decision-making. To gain stakeholder buy-in, IBM employs a champion system where employees advocate for technology adoption within their domains. Additionally, it implements citizen technologist programs to empower employees with the necessary skills to leverage new technologies effectively, which is about establishing AI creators and enabling them to train, test and deploy AI within their domain. Those technologists are not only actively implementing AI into workflows but also ensuring the adoption of AI within the teams.

Change management training and support to complement training of users is important since having an appreciation of the significance of change can help build understanding and empathy between the project team and the business users. Conducting 'impact' assessments helps the project team understand the change. It helps inform training requirements, which should always provide a special, unique experience for the users: don't take the vendor documents and hope that is the answer. Also, as we saw before, it is a nice touch to have sponsors and the leadership team includes videos and messages explaining why the company is making this change and updating these messages throughout the life of the project; it resonates much stronger than hearing the same messages all the time and helps employees understand the progress made.

The suggestion that post-project, the new supply chain solution is embedded so change management training and support can end, is a failing of many projects, because often, the relief of finishing overtakes the importance of the continuous learning.

The ongoing training requirements should be handed over to the leadership of the function that is using the new system. These leaders should work with HR to keep the materials refreshed and challenged. Metrics should be developed to understand engagement with the new solution and tracked for at least the first six months post 'go live' to ensure the system is being used in a way that is aligned to leadership's expectations and that users haven't continued with their old ways of working and bypassing the solution. Using forums and surveys to engage users will help provide qualitative input aligned to quantitative data from metrics. Over time, this should address any concerns and improve adherence to its use. Lessons learnt should be completed through the life of a project at key milestones (as already mentioned in Chapter 9) and at the end of the project before the project team disbands. This documentation should be available so that the company employees can refer to it for future implementations.

One further consideration is cultural. International projects can involve overcoming different forms of communication and ways that it can be wrongly interpreted.[23] British phrases can be interpreted in different ways and leave them open to misunderstanding. For example, someone saying 'that is fine' really means 'could have been better' but the non-native takes away the words literally. From verbal to written forms, this topic can cause lots of mistrust unintentionally, so time must be spent understanding and respecting different cultures when working in project environments. There are many times when external parties are put together quickly to form a team. This reinforces the need to spend time in each other's company to build rapport, trust and understanding.

EXPERT INSIGHTS

On discussing change management with an experienced operator who has worked with a variety of pharmaceutical and consumer goods companies, what is clear is that organizations often overlook cultural implications, which can encourage resistance and suboptimal outcomes. Therefore, early stakeholder involvement to foster advocacy for change and the need for flexibility in approach that adapts to the urgency of the situation is important. Effective communication of the benefits of change needs to be front and centre when formalizing the project and in updates throughout the life cycle of the project. These point to a deliberate strategy for communications that assist leaders to gain support and mitigate detractors' influence. This is where external perspectives can add value to the internally led 'change' team, providing objectivity and clarity with their experience in doing similar initiatives with other companies.

On culture, the approach to change must consider the style the company operates under, whether it is command and control or a collaborative approach. While most companies will claim to have the latter, many successful supply chains operate using predominantly the former approach. Therefore, how the change is driven by leadership may need to adapt. Also, the area the technology is being implemented within is important to clarify because while supply chain professionals may sit organizationally under the same leadership, their motivations and different mindsets need to be understood – it should not be assumed that how they will react and engage with the change will be the same.

A supply chain technology implementation always should come back to being transparent, consistent and willing to be open with the project team and users, so they and the company gain the most from the new solution. Yes, with any change, there is naturally always an element of fear. However, establishing a model or framework to use throughout the transformation will help to alleviate most of those concerns and provide the best opportunity for success.

There are a variety of these models and frameworks available, which have been adopted over time to assist companies through the change. For example, Kotter's 8-step change model[24] has been successfully implemented by companies looking at change at both macro (company-wide) and micro (function) levels.

The steps of Kotter's model are:

1. Create a sense of urgency.
2. Build a guiding coalition.
3. Form a strategic vision.
4. Enlist a volunteer army.
5. Enable action by removing barriers.

6 Generate short-term wins.

7 Sustain acceleration.

8 Institute change.

For example, under the stewardship of Alan Mulally, Ford was turned around in a way that is still revered today. He initiated change at Ford by establishing urgency and formed a coalition of top executives to shift the company's focus from large, gas-guzzling vehicles to fuel-efficient ones. This vision was communicated through town hall meetings, empowering employees by removing barriers and providing resources (e.g. funding for initiatives and additional people such as change managers). Celebrating short-term wins, like the successful launch of the Ford Fusion, helped build momentum. Over time, these changes became embedded in Ford's culture, making the company more agile and responsive to market changes, ultimately returning to profitability and avoiding bankruptcy during the 2008 financial crisis.[25]

However, this approach is somewhat linear, does not flex well and the leadership approach is command and control (top-down). Company employees may not respond well to that if there is a culture where innovation is driven bottom-up, for example,[26] so the model needs to be carefully considered. Other models are available such as the Prosci ADKAR model,[27] which is more about making change stick by assisting users through the change (bottom-up).[28]

Whichever model a company uses, the following points should be considered from the outset to the conclusion of the project.

Create urgency and desire for the people of the business to get behind the solution and the team delivering it. Leadership needs to help create an environment for the project to prosper and help balance the day-to-day business with delivering the project at pace. By doing a stakeholder analysis early, the project leadership team can understand the level of support they have throughout the company and where to focus time to get buy-in or hear objections proactively and address them. I have had experience of needing to do this with non-executives who had cold feet in relation to the implementation of an inventory optimization solution with a retailer. Through a couple of working sessions explaining and demonstrating the solution and how the solution compared to what their competitors were doing, they understood the benefits and moved from detractors to supporters, which helped with positioning other C-suite participants to approve funding.

Develop a change management plan of activities and dedicate appropriate resource commitments throughout the project. This is so employees know their roles and no assumptions are made in relation to expectations of double-hatting between their day job and the project, unless required and if so, be clear on the expectation of time. This may have to change over the course of the project; however, documentation should be kept updated to record this and agreed with the employees and their managers. The number of times I have seen change management as a negotiable is astounding; however, the value it brings when done right is immeasurable. Why it sometimes is a struggle is in the delivery of the activities; we can sometimes focus too much on just getting them completed. Also, it must be ensured that the people delivering the change understand the solution and the roles affected, otherwise there can be a disconnect and a lack of trust between the function and them.

Looking to the future

Many of the strategies seem like common sense, yet the gaps still seen between supply chain technology implementations that excel and those that don't show that some companies have a way to go in adopting change management.

Change management is a term that will not disappear when it comes to embedding new supply chain technology in companies. What will be different is the tools we use to help deliver projects – as we saw, for example, during Covid, whereby we started using video conferencing as modus operandi for business. We became comfortable staring at a shared space online as we developed and discussed process maps to user journeys to IT architecture and beyond. Also, PMO and project collaboration tools have advanced, which makes these far more automated and less time-consuming to use.

Supply chain success requires people, process, technology and data to work in harmony. No technology change should be considered in isolation of the other elements or it is doomed to fail. Whatever the future holds, we need to be cognisant that success lies with enterprise change.

From a cultural transformation perspective, the supply chain has changed and continues to. Global collaboration is now a must for most companies and to operate this way requires them to confidently use technology, linking it with partners upstream and downstream. The adoption of new supply chain solutions is a necessity in strategic planning for companies that want to succeed in the long term. For many years, investments into core systems have been put off, but the pace of technology advancements means not embracing change and investing in it could lead to dire consequences. There needs to be rigour and accountability through the life cycle of the project from its inception, but be mindful of what is the right level for your company because if you have a high-performing culture, putting obstacles in the way will disengage all participants quickly.[29]

Conclusion

This chapter has most likely felt like it is applicable for any project and that has been my intention. People always say their company is very different from others but when it comes to implementing technology solutions, the reality is that they have far more in common with other companies than they will admit or realize.

New technology in supply chain management is a complex process that requires considered planning, execution with urgency and ongoing support. People are at the centre of this, so the effectiveness of the technology will stand or fail based on how successful the company has been in equipping them with training and development and whether they have used this opportunity to reimagine how workflows should be done.

Projects can struggle at various times. In my experience, this is usually in the 'design' and 'testing' phases, when what is built goes through rigorous challenge during the journey of connecting to a company's infrastructure and is exposed to user engagement.

Supporting people to be able to go through this change requires effort and consideration. Engaging with different groups of people, whether stakeholders or users, is crucial and should be treated that way with consistency and transparency. The best practices and suggestions have been discussed and these should be followed and customized to be appropriate for what works for your company and culture. It is too easy to go with the vendor's generic approach and force it into the company; however, those responsible for the change should appreciate and understand what resonates with their people. The key takeaway is that the users should be encouraged to engage and participate throughout the process because following this approach will result in ownership by them to take this change forward as business as usual.

The importance of change management and cultural transformation should not be underestimated; they are still not considered as a priority for delivering successful supply chain technology. We get fixated on the technology or getting through the next milestone that we devalue them as tick-the-box activities.

They are a cost, where sometimes the link to their benefits isn't as clear-cut as improved forecast accuracy or reduced cost to serve. Any technology project in the supply chain should always take time to understand what capabilities the solution is providing, as well as how the users will go through the change journey, preparing them to use the solution and realize the 'perceived' benefits. Supply chain leaders have not done themselves many favours by holding onto manual ways of working with bygone software, which are unhelpful to them with the global nature of many companies today.[30]

The fact that implementing technology can have significant ramifications on a company's bottom line only reinforces the need for taking these types of projects seriously.

Many of the reasons why companies don't deliver comes down to change and culture readiness. Were the users prepared for the change? Were they excited by it? Could they see the benefits this solution would provide them in their role? In many instances, the answer is no, but to be successful with any transformation, we must overcome these obstacles and by following the strategies in this chapter, your ability to lead (formally or not) and deliver projects that people want to be involved with will increase and result in better outcomes than those that don't.

The success of technology implementation and adoption relies on the people. Through prioritizing training and development, change management and intentional communication, companies will be more likely to have successful transitions to new technology capabilities.

Endnotes

1 Goodell, J (2011) Steve Jobs in 1994: The Rolling Stone interview, *Rolling Stone*, www.rollingstone.com/culture/culture-news/steve-jobs-in-1994-the-rolling-stone-interview-231132 (archived at https://perma.cc/6MJ3-MMQU)

2 Dolfing, H (2020) Case study 12: Lidl's €500 million SAP debacle, henricodolfing, www.henricodolfing.com/2020/05/case-study-lidl-sap-debacle.html (archived at https://perma.cc/77AC-9HJ2)

3 Consultancy.uk (2018) Lidl cancels SAP introduction having sunk €500 million into it, Consultancy.uk, www.consultancy.uk/news/18243/lidl-cancels-sap-introduction-having-sunk-500-million-into-it (archived at https://perma.cc/NZW9-UUL7)

4 Roberts, R, Flin, R, Millar, D and Corradi, L (2021) Psychological factors influencing technology adoption: A case study from the oil and gas industry, *Technovation*, 102, www.sciencedirect.com/science/article/pii/S0166497220300912 (archived at https://perma.cc/G3NY-BFAA)

5 Brooks, G (2025) Breaking down barriers to technology adoption in global supply chains, *Global Trade Magazine*, www.globaltrademag.com/breaking-down-barriers-to-technology-adoption-in-global-supply-chains (archived at https://perma.cc/B7UT-3KNG)

6 CEO Weekly Staff (2025) Overcoming barriers to tech adoption in global supply chains, *CEO Weekly*, https://ceoweekly.com/overcoming-barriers-to-tech-adoption-in-global-supply-chains (archived at https://perma.cc/V2BF-9ZDA)

7 Vulfsons, E (2023) How tech-driven change management boosts employee development, *Training Magazine*, https://trainingmag.com/how-tech-driven-change-management-boosts-employee-development (archived at https://perma.cc/J729-RA92)

8 Detro, J D (2024) Examining the impact of supply chain technology implementations on supply chain effectiveness and firm value, *Global Journal of Business and Integral Security*, www.researchgate.net/publication/388855865_Examining_the_Impact_of_Supply_Chain_Technology_Implementations_on_Supply_Chain_Effectiveness_and_Firm_Value (archived at https://perma.cc/L9EF-TQS4)

9 Shah, H S and Bohlen, J (2023) Implicit bias, National Library of Medicine, www.ncbi.nlm.nih.gov/books/NBK589697/#:~:text=With%20explicit%20bias%2C%20individuals%20are,clinical%20interactions%20and%20healthcare%20delivery (archived at https://perma.cc/U53D-XEBS)

10 Grupe, D W and Nitschke, J B (2013) Uncertainty and anticipation in anxiety: an integrated neurobiological and psychological perspective, *Nature Reviews Neuroscience*, www.nature.com/articles/nrn3524 (archived at https://perma.cc/AWM2-NPGF)

11 Alicke, K, Dumitrescu, E and Protopappa-Sieke, M (2022) Transforming supply chains: Do you have the skills to accelerate your capabilities? McKinsey & Company, www.mckinsey.com/capabilities/operations/our-insights/transforming-supply-chains-do-you-have-the-skills-to-accelerate-your-capabilities (archived at https://perma.cc/Y33W-FXSL)

12 ActivTrak (2023) How can technology improve productivity? ActivTrak, www.activtrak.com/blog/how-can-technology-improve-productivity/#:~:text=Improved%20communication%20and%20collaboration%20tools&text=Technology%20can%20help%20employees%20stay,bottlenecks%20that%20create%20timeline%20delays (archived at https://perma.cc/7MC4-NQJE)

13 Dennis, A (2024) What is a sandbox environment? Benefits, use cases, whatfix, https://whatfix.com/blog/sandbox-environment/#:~:text=Sandboxes%20provide%20a%20safe%20space,fix%20bugs%2C%20and%20validate%20changes (archived at https://perma.cc/ANQ6-N8RD)

14 Ticong, L (2024) Why is data management important? 8 key benefits, Datamation, www.datamation.com/big-data/why-is-data-management-important/#:~:text=Effective%20data%20management%20supports%20decision,increased%20customer%20satisfaction%20and%20competitiveness (archived at https://perma.cc/FZ55-8QMH)

15 Nicolas, J (2020) Day in the life technique | Get a clear perspective of your users, Business Analyst Mentor, https://businessanalystmentor.com/day-in-the-life-technique (archived at https://perma.cc/KC62-XATN)

16 Morukian, M (2023) Breaking down 'us vs. them' thinking, Chief Talent Officer, www.chieftalentofficer.co/2023/03/22/breaking-down-us-vs-them-thinking/#:~:text=When%20we%20are%20in%20the,feel%20comfortable%2C%20secure%20and%20understood (archived at https://perma.cc/3Z2P-WVL9)

17 Immel, J (2019) Why is train the trainer in software implementation projects an effective exercise, Stoneridge Software, https://stoneridgesoftware.com/why-is-train-the-trainer-in-software-implementation-projects-an-effective-exercise/#:~:text=What%20is%20"Train%20the%20Trainer,attached%20to%20that%20person%20leaving (archived at https://perma.cc/Y58M-JGNG)

18 Mutambik, I (2024) The role of strategic partnerships and digital transformation in enhancing supply chain agility and performance, *Special Issue: Blockchain Technology in Supply Chain Management and Logistics*, www.mdpi.com/2079-8954/12/11/456 (archived at https://perma.cc/DB63-E8VB)

19 Continu Team (2022) Engaging with the 4 different types of learning styles, Continu, www.continu.com/blog/learning-styles#:~:text=The%20four%20learning%20styles%20are,learning%20through%20doing%20and%20experiencing) (archived at https://perma.cc/F4HX-V79F)

20 Morganelli, M (2023) Types of learning styles: The value in holistic strategies, Southern New Hampshire University, www.snhu.edu/about-us/newsroom/education/types-of-learning-styles#:~:text=While%20the%20VARK%20learning%20styles,as%20the%20most%20common%20combination (archived at https://perma.cc/6S6W-5AJT)

21 Grover, S (2022) UX strategy for supply chain design, Onething Design, www.onething.design/post/ux-strategy-for-supply-chain-design#:~:text=1.,and%20improve%20your%20UX%20design (archived at https://perma.cc/2DL2-6NRZ)

22 enov8 (2024) Mastering the art of cutover planning: A step-by-step guide, enov8, www.enov8.com/blog/mastering-the-art-of-cutover-planning-a-step-by-step-guide (archived at https://perma.cc/Q7JW-RJNT)

23 Livermore, D (2016) Weird, rude, or different?! Awkward cross-cultural moments, Cultural Intelligence Center, https://culturalq.com/blog/rude (archived at https://perma.cc/4UFJ-C6PM)

24 Kotter, J P (2008) *A Sense of Urgency*, Harvard Business Press

25 Cable, J (2011) Management lessons you can learn from Alan Mulally, *Industry Week*, www.industryweek.com/the-economy/article/21953589/management-lessons-you-can-learn-from-alan-mulally (archived at https://perma.cc/LKP8-YFN5)

26 Olmstead, L (2019) Kotter's 8-step change model: Advantages & disadvantages, Whatfix, https://whatfix.com/blog/kotters-8-step-change-model (archived at https://perma.cc/K92G-N5FK)

27 Ball, K (2024) *The ADKAR Advantage: Your New Lens For Successful Change*, Prosci

28 Brusati, I (2025) ADKAR vs Kotter: Which change model should you choose? Prosci, www.prosci.com/blog/adkar-vs-kotter (archived at https://perma.cc/LLH8-9XUJ)

29 Kotter, J P, Akhtar, V and Gupta, G (2021) Is your organization surviving change – or thriving in it? *Harvard Business Review*, https://hbr.org/2021/08/is-your-organization-surviving-change-or-thriving-in-it (archived at https://perma.cc/ZE3B-3VZJ)

30 Destino, M, Fischer, J, Müllerklein, D and Trautwein, V (2022) To improve your supply chain, modernize your supply-chain IT, McKinsey & Company, www.mckinsey.com/capabilities/operations/our-insights/to-improve-your-supply-chain-modernize-your-supply-chain-it (archived at https://perma.cc/M9TH-7P37)

Succeeding in supply chain technology implementation

12

This chapter will discuss how to succeed in supply chain technology implementation, including integration challenges and data quality issues, and how to overcome these – from planning, to preparation, to implementation.

CHAPTER AIMS

- Anticipate common challenges in supply chain technology implementation.
- Learn how to overcome through planning, preparation and implementation.

'In the middle of every difficulty lies opportunity.'

WIDELY ATTRIBUTED TO ALBERT EINSTEIN

The implementation of supply chain technology can be a complex and challenging process. Despite the desirable potential benefits in many instances, companies can struggle to overcome obstacles and avoid pitfalls that can knock projects off their path, leading to delays or in extreme cases, abandonment.

If we consolidate our thinking over the last few chapters, as well as introduce other elements in this one, we can overcome those obstacles and hazards and successfully implement supply chain technology solutions and keep them relevant. These obstacles are outlined in Figure 12.1.

Figure 12.1 Obstacles to successful supply chain technology implementations

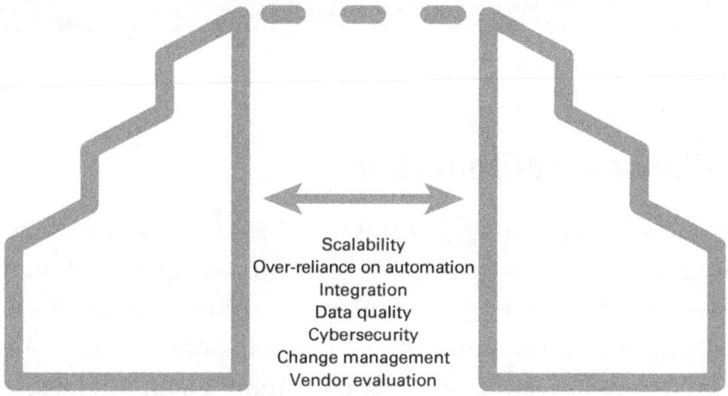

Common obstacles and pitfalls

There are numerous obstacles and pitfalls that are often encountered when implementing supply chain technology, including scalability, over-reliance on automation, integration, cybersecurity, change management and vendor evaluation.

Scalability

The scalability of the chosen technology can cause significant challenge. This is because looking at technology in isolation means that decisions are being made based on a host of assumptions that do not appreciate that variables could be different for different markets that the company operates in (e.g. data is surfaced in a standardized way for recording, but each market might use different attributes). These assumptions may be sound and reasonable at the time but may be found not to be so credible in the future. Also, a company's operations can expand, leading to further complexity as the solution then must accommodate for increased demand, transactions and data. Furthermore, solutions take time to be implemented and the operating model, processes and interactions with third parties upstream and downstream could make them redundant when brought online. The complexity of operating globally can make it difficult to accommodate users' needs in diverse geographies and business environments. Plus, there can be local nuances that are non-negotiable, which only come to light when you are focused on a local deployment and discuss their requirements beyond the global template developed (e.g. tariffs, different languages, currency, how they use numbers and view data).

Over-reliance on automation

With new technology in supply chains, the theme of automation is usually emphasized as one of the key reasons for why companies should invest. With that in mind, becoming over reliant on automation can have significant ramifications. Automation should provide opportunities to improve efficiency, but it can lead to making the supply chain vulnerable. There can be a loss of control because over-relying on automation reduces user oversight, which could lead to many mistakes going through the systems unnoticed until significant errors occur. Furthermore, automation may struggle to consider factors beyond data and respond to unexpected events such as supply disruptions and demand fluctuations. During the Covid pandemic, some companies found themselves having to spend considerable time overwriting their supply chain systems to bring forecasted demand down. If they hadn't done this, they would have flooded too much inventory into the market when consumers were having to adjust to purchasing primarily online. For example, the US retailer Target had a multibillion-dollar inventory challenge where it couldn't move inventory fast enough due to the disruption. It was struggling with matching inventory (which could be on long lead times from China in some instances) and trying to serve from a vast store and distribution centre network. This involved finding ways to convert stores into online fulfilment and cutting back demand on systems. It took Target two years post-pandemic to get inventory stripped back by $2 billion (~14% reduction).[1]

Integration

Integration poses significant challenges, particularly in relation to compatibility, data formatting and mapping and application programme interfaces (APIs).

System compatibility is one of the most frequently cited integration issues companies face when embedding new supply technology in their IT landscape.[2] The process of connecting software and hardware modules into one complete picture can be difficult and time-consuming. What can make this harder is that a company's IT architecture can always be in flux. Especially in larger companies, there is always tinkering going on.[3]

For example, in 2010, BP had 90 separate instances (i.e. different versions) of SAP for ERP functionality, with a mission to implement modern versions of its ERP solution and simplify the number of instances to 40.[4] A new supply chain technology capability may have to integrate to many different ERPs. This can be made even more challenging when these different ERPs may be upgraded to fewer versions at the same time as implementing the new system. You can quickly get the feeling post-implementation that you are walking around with a plug trying to find the correct socket to connect it to.

Once you are comfortable with system integration, you then need to look at how the processes align for the new technology as part of the overarching supply chain orchestration. This can involve reviewing processes at various levels aligned to a framework such as the Supply Chain Operations Reference (SCOR) model[5] and making sure they are documented and maintained. SCOR provides a somewhat standardized framework for supply chain processes (e.g. plan, source, transform) and is applicable across various sectors, providing guidelines for companies to then develop their own process taxonomy.[6] It focuses on processes and does not align to subfunctions or roles and people. Level 1 is the process type (the strategic level). Level 2 is where each process is broken down into process categories (the tactical level). Level 3 is where the processes are broken up, identifying the strategic element within each process (the operational level). Level 4 is where each strategic element is broken up, creating a flow of tasks and activities for each element (the detailed level).

As an example,[7] if we were wanting to document processes for fulfilment (the stages of the delivery process), it could look like:

- **Level 1** – Fulfil.
- **Level 2** – Fulfil business to consumer / fulfil business to business / fulfil intra-company.
- **Level 3** – Following fulfil business to consumer only. The sequence could include:
  - receive business to consumer product from source or transform (i.e. manufacturing)
  - receive order signal
  - pick product

- pack product
- stage product
- schedule transportation
- notify and confirm shipment window
- load vehicle and generate shipping document
- ship product
- assemble or install product
- obtain proof of delivery or customer acceptance.

- **Level 4** – Document the tasks for each level 3 process in detail. The process could include:
  - detailed instructions for each level 3 process step by step on how to perform specific tasks
  - system instructions outline the inputs, outputs and rules associated with each step
  - IT specifications detailing technical requirements
  - training and standard operating procedures so that any employees understand the processes but also use them in a consistent manner.

However, do note that it is for each company to look at their uniqueness and document appropriately, rather than take the SCOR model as fact and try to fit your company into its framework. I find that using level 1 of the model and then reviewing the other levels with supply chain teams, understanding what they do compared to the model, is a solid approach to making sure the key processes to build the taxonomy are covered.

Integration then can move beyond the company's four walls, with further integration with customers (downstream), suppliers (upstream) and logistic providers (both upstream / downstream), which can cause potential disruption in the end-to-end supply chain. For example, when Flora Food Group was establishing itself as a new entity and implementing new capabilities, it had a challenge getting logistics providers set up to receive and send messages of shipment requirements to be acted on. This required orchestrating different companies throughout testing to make sure all parties were receiving correct information, which became a time-consuming activity. This was crucial to complete due to the revenue associated with a significant few customers (i.e. US Grocers) as the receiver of its products.

Data formatting

Integrating systems quickly brings us to the issue of amalgamating data from different systems and formats. Depending on the solution chosen and the systems being integrated, data formatting and mapping issues can occur. To overcome problems with different data formats interacting across systems, it is always wise to have a 'parent–child' relationship with data. This is a hierarchical relationship where the 'parent' serves as the master reference point and the 'child' hangs off it for key information – think of a stock-keeping unit and having the same code across multiple systems.[8] The 'child' may include other details but keeps the code of the 'parent', which is usually found in a dedicated master data management system or stored in the ERP system.[9]

While this sounds logical when written down, it is only when you start to investigate the data and try to map it between systems that issues can appear. Translating fields of information and then seeing how they map across can be time-consuming and lead to errors, especially during testing. These errors can sometimes go hidden until the technology is in the users' hands. However, depending on the types of testing done (e.g. system integration testing, unit testing, compatibility testing, functional testing, acceptance testing), this should provide enough viewing of outputs, and having clear criteria for what is deemed pass / fail should help minimize impact. For example, for unit tests, does the test execute the task requested in a specific part of the system? Yes or no means it has passed or failed respectively.

The mindset should be to flush out all problems during these activities. I get nervous when testing goes smoothly because the worry is, has something been missed? Integrating a new technology should allow time for it to be run, see whether it passes or fails according to the criteria, is updated and run again on a loop until it passes. This can lead to hours of re-work and is why using the sprint approach[10] (a core component of the agile approach) to tackle this is best, so any errors can be picked up as backlogs and prioritized for the next sprint. Not doing this could delay the project moving to the next phase if following a traditional waterfall approach (whereby each phase has to be completed before moving forward).

Creating new data

Once the data mapping is done, understanding what new data must be created can also be a gap that is not seen early enough. Most software companies should be able to highlight where there will be gaps; however,

I was asked recently for a view on a project where levels of planning were absent. They were forecasting at an aggregate level of category rather than by SKU from the data that had migrated from the old system to the new one. This meant they had to try to plan demand at an aggregated level and build up a forecast in near real time at SKU level, while the formatting and mapping of data was hastily done post go-live.

APIs

APIs, which we discussed in Chapter 7, are another integration point that can have connectivity issues. One of the key steps to overcome this involves the programmer developing an API specification. This specification should outline what systems are being connected via APIs and what is happening between the systems when updates are done to one. APIs are popular for connecting with outside organizations and if done correctly, only expose the objects or actions needed to operate. For example, if your business is showing traffic via an app, by having sensors that offer an API in different locations, the app can send messages (a call) to feedback reports (response) of traffic conditions that are automatically updated. If any of those sensors start to have issues feeding back data for whatever reason, then they can become a liability and could lead to users getting stuck in traffic. As an extreme consequence, complaints about your app could lead to you going out of business. The same can be said for other systems. Therefore, the risk is if there is a break in the connection between the API call and response, then systems will not be updated or updated incorrectly, leading to issues.

Data quality

Data quality issues continue to make integration difficult and can stop a project in its tracks. Often, there is the fear of exposing data and demonstrating how poorly governed and appreciated it is within a company. Data quality provides a competitive advantage and needs to be treated as an asset to the business. Data quality issues fall under three headings: inaccurate and/or incomplete data, isolated data (i.e. duplication in various systems) and governance, which we will explore now.

Inaccurate or incomplete data is a telltale sign of wider issues. Not having strong data quality will lead to decisions being made that lack insights and can have catastrophic outcomes. For example, it can lead to making decisions about inventory and buying that are out of sync with demand. This can have significant material ramifications, such as having far too much inventory on hand that then has an impact on inventory costs, storage costs, shelf life and rendering products redundant (dependent on the products your company buys, transforms and sells).

Fragmentation of data, kept in silos within a plethora of systems and functions of the business, will make it difficult and time-consuming to build a picture of current information to then make decisions from. It could lead to inaccurate data driving the company's supply chain and beyond and comes back to the point of not having a single source of truth. Decisions that companies make in relation to parent and child data are critical here because users could be accessing information that they believe to be correct, when in fact, looking at another system might reveal a different answer. For example, if you have not defined the relationship, you could have the same product in different systems with different reference codes and descriptions (plus other attributes), which then makes it hard to keep control of how well the product is being managed through the supply chain. You are then unable to marry the demand and supply sides.

Governance of data and who is responsible for it can lead to issues. Usually, it makes sense for IT to set the guiding principles but keeping data updated is the responsibility of the functions and subfunctions of supply chain, in my opinion. For example, when a new finished good is created, the bill of materials (BoM) should be inputted as the master into the 'parent' system (typically the ERP), which then feeds into other systems so that buying and supply planners can then act on it. Keeping it updated, then, should fall to the buyers, who will input commercial data for raw materials and packaging. Supply planners are responsible for refreshing the constraints of the material buy such as minimum / maximum order from discussions with buyers. Not having a standardized approach for governance will lead to inconsistencies and errors, which may go unseen until a significant issue happens. Also, as important as keeping data is for 'live' products, it is as important to have rules in place about how to make products 'inactive' so that no purchases are made against these by accident.

Cybersecurity

With the adoption and embracement of digital and connected supply chains, the risks of cybersecurity have become more prevalent. By having the supply chain connected with IoT devices, cloud platforms and integrations across the globe, both internally and externally, the company could be exposed to cyberattacks and data breaches. For example, in late 2024, Blue Yonder suffered such a cyberattack leading to severe consequences for a few of its customers, including Starbucks and Morrisons, which severely disrupted logistics operations.[11] Companies should consider their vulnerability to data leaks, which can then be exploited by hackers and surface externally sensitive company information. This is where having a register for all systems in use across a company and evaluating it regularly for adherence is a must. To lessen your vulnerability, the technology solution for the supply chain (as well as the rest of your systems) should have the appropriate security features to protect the company. Also, ensure that the parties that systems connect to externally have the appropriate levels of security as well, which should be enforced as part of contracts and audits with vendors.

Change management

The importance of change management was highlighted in Chapter 11, and as was discussed, this can be one of the main stumbling blocks for companies. It is worth stressing again that implementing supply chain technology requires significant changes to business processes and user behaviour; it should never be considered in isolation. Unless there is a necessity to implement a new technology due to the age of current systems or compelling benefits, I would question why you are doing it.

Common change management issues include resistance to change, which when not tackled by the company, will lead to poor adoption and can make users less productive as they try to retrofit their ways of working on a new system. Lack of time and investment for training and support will drive frustration further. Many of these types of implementations are deemed to be IT-driven and this view needs to change. Yes, the IT team may take the lead on implementing the system but there will be limited benefits if the supply chain function is not onboard and engaged to reimagine their processes to get the best from the solution. Finally, any lack of communication between stakeholders (internal or external) can quickly lead to misunderstandings and delays. The success of a project can at times be attributed to communication above all else; doing it well will lead to trust and motivation from all parties to stay the course and contribute.

Vendor evaluation

The choice of vendor for the solution can cause challenges for companies that are swayed by the 'can do it all' approach. Managing the relationship effectively is critical for the success of the implementation. Time needs to be spent early on, understanding the solution from business and IT perspectives, having demonstrations and learning from the vendor about how they have implemented this with others. Jumping too quickly to securing a vendor can result in the solution not meeting the business needs, which drives compromises early into the process unnecessarily.

Contracting with vendors can result in hidden costs and decreased value if not considered from the view of the overall project. Numerous times I have seen the squeezing of vendors on cost, which might result in a short-term win for the company, but long-term increases due to the need to purchase licences and a compromised IT landscape, for example. One well-known vendor offers free licences for some of its complimentary solutions to appeal to procurement. However, this can lead to a mismatch in expectations when the solution does not fit what the business requires and it is unable to go in a different direction, being handcuffed when inevitably upgrades are required.

Checking on what support the vendor offers should be prioritized early on, since if the solution is new to market, there may not be a mature market of integrators and contractors. Therefore, this poses a risk to the company if it is solely reliant on vendor resources. The risk can be quantified in relation to the time it takes to resolve issues – from identification to user satisfaction with the solution. Making sure that the company representatives stay in the driving seat is important because at this stage, there is limited trust and the vendor will not know the people and the company culture.

What to consider

So far in this chapter, we have surfaced a variety of issues to watch out for during the life cycle of implementing a supply chain technology solution. Now we turn our attention to activities that should be done to reduce the risk of these becoming issues for you.

Scalability

Scalability can be a 'Catch 22' situation since you can spend so much time on it that you may never decide. Leveraging cloud-based solutions offers the company better scaling options and should be a must in most instances. This may not be the case in relation to some defence situations, for example, where they may demand on-premise (runs on computer servers at the location used rather than on cloud). Going with a modular solution where additions and upgrades can be considered at later times will help and may reduce the burden on users, who can get familiar with a new solution first and then, in time, add extra features.

EXPERT INSIGHTS

As you enter the design phase, the importance of sticking to the initial business case and avoiding scope creep is critical. Any additional ideas captured should be set aside for future phases to maintain team focus on core objectives. When requirements should be completed is a difficult activity to determine. Ideally, they should have been agreed prior to commencing the design phase but they sometimes linger on. The lesson here is to innovate during the mapping of requirements, when you decide whether the asks are 'must have' or 'nice to have', then agree and sign off so to avoid costly changes later before the design phase. Signing off should be formally completed with all relevant sponsors and stakeholders. The number of times stakeholders and sponsors agree to a design, then when something occurs later, try to evade responsibility, is quite surprising. Vendor resources are critical, particularly in this phase: ensure all is signed off between each vendor.

Also do not accept any customizations. Reporting is an area that should use your company's standard business / management information tools. Don't create custom reports in the solution to go live with the project, do that once the system is stable, if necessary.

Scalability is a crucial consideration for technology investments. Organizations must ensure that their systems are adaptable and can grow with the business. The feasibility of global solutions versus local adaptations is a key question in a rapidly changing technological landscape. It is essential to have systems that are flexible and can be tailored to meet the unique needs of different regions and markets.

Over-reliance on automation

To mitigate the over-reliance on automation when choosing a supply chain solution, understanding what the solution can do automatically and the ability to switch capabilities on/off and explain output are key. Oversight and decision-making should be balanced more towards the user, with them having the ability to intervene where necessary. Where that is not an option, contingency plans should be developed in case you need to adapt quickly to an event.

In the early stages of planning, an exercise should be conducted with supply chain users to document their requirements of the new solution, looking at it firstly from an unconstrained view and then reviewing with them in separate sessions to challenge which are 'must haves' versus 'nice to haves' versus 'not needed'. It can be helpful here to also bring in external perspectives to help the users to consider additional requirements that they may not have considered based on their experience being primarily on the current system and ways of working. The importance of doing this activity should not be underestimated because it helps those representing the supply chain within the company confidently know what they are looking for when speaking with vendors. This will help the vendors to understand quickly whether their solution is a fit or a partial fit and if the latter, what requirements cannot be delivered.

For example, if you are looking at implementing a warehouse management system (WMS), by working with teams in the warehouse, distribution planning, customer service, manufacturing and relevant transport providers, you can quickly begin to determine what you need the new system to do. You can also take forward key business requirements from the current system. Some of the business requirements I would expect to see include:

- Create a pick plan by picking type, matching products to specific orders and showing order status.
- Create a shipment order when a connected system (e.g. order management system) signals demand.
- Create putaway algorithms, determining suitable location based on type, size, volatility and velocity.
- Run cycle counting, create an audit of when products were last touched, such as time, user, last movement and inventory levels.
- Track items by batch or lot, notifying users of quality issues and then place products on unavailable (e.g. reject, hold).
- Cross-dock logic to allow products to move through the warehouse from inbound to outbound bypassing storage.
- Handle returns and guide this based on user input to a suitable location in the warehouse to assess, scrap or repair and recycle.
- Ability to generate reports including activity levels, productivity, order cycle times, storage density, shipment and inventory accuracy, cycle counting.

Some readers may argue that this should be the first step of the design phase, rather than being done prior to engagement with vendors. However, having an initial list of requirements earlier helps with building the collaboration between teams and helps to accelerate vendor selection. During the design phase, you will build on these to formalize them, then develop functional requirements and technical requirements before moving on to solution design, architecture, process design and mapping.

From the engagement with vendors, implementation plans should be co-created (you might bring on support from the system and/or business integrator as well at this time) at the milestone level, with timelines and indicative resource allocations, which can be refined further as you reiterate the business case.

Integration

Developing an integration strategy for the company should go beyond just looking at the new supply chain technology solution specifically, so when you add new capabilities, the integration process is clear.

However, in relation to the new supply chain technology solution, the key considerations include defining data formats and how they map together, and the governance to oversee the execution. Identifying early in the process the number of interfaces required can save you from issues such as delays later. For example, the time required to implement an interface can be a few weeks to a few months, depending on the complexity of the systems interfacing between. It is also helpful here to lean on the system integrator and vendor solution provider chosen because they should have the experience of knowing what has worked with other similar companies and the effort required to do the work.

During the testing phase, the testing and validation of interfaces (integration between specific components) and integration (testing of multiple components to make sure the solution as a whole works) are completed. In most instances, this is done by a system integrator. When it comes to user acceptance testing (UAT) (i.e. the testing completed with the users of the system in the future), scripts need to be effective, cover the key processes that will be executed in the new system (e.g. WMS would include goods receipt, putaway and storage, picking, packing, shipping, internal transfers, cycle counting) to identify and rectify any differences between the intended design and the user experience. In preparation of this (i.e. during the early test phase), the project team should understand the requirements and have the design beside them so they can develop scripts and then review with the business users of the new system in advance of this stage of testing (remember the testing phase includes a number of different tests from unit, to system, to integration, to UAT). How much testing to do is a tricky question to answer, but typically, you would expect to see for most supply chain technology solutions, a timeline of 4–6 weeks. Time needs to be given to review and address any tests that fail as well as the tests need to cover a sample of different users within the business who will engage with the system (this also helps with getting users engaged with the system). Table 12.1 shows a few examples of UAT scripts for WMS.

Table 12.1 UAT scripts for WMS

	Inventory management	Receiving and putaway	Shipping and picking
Title	Create new product	Create order, receive and putaway	Create shipping order, pick and pack and ship
Steps	• Navigate to inventory module • Search for a specific product by name or SKU • Verify product details (e.g. quantity, location, status) • Update product quantity and verify changes • Create a new product and verify successful creation	• Navigate to receiving module • Create a new receiving order and verify successful creation • Receive a shipment and verify successful receipt • Put away received inventory and verify successful putaway	• Navigate to shipping module • Create a new shipping order and verify successful creation • Pick and pack order and verify successful pick and pack • Ship order and verify successful shipment
Expected results	• Product search returns correct results • Product details are accurate and current • Quantity updates are successful and reflected in the system • New product creation is successful	• Receiving order creation is successful • Shipment receipt is received and inventory is updated • Putaway completed and inventory is updated	• Shipping order creation is successful • Pick and pack completed and inventory is updated • Shipment is executed and order is marked as shipped

EXPERT INSIGHTS

People underestimate the effort and budget needed to set up different environments and to do testing. On integration testing, ensure it is done across systems and not in isolation (i.e. test to the end of one system, then test the second system). Testing needs to see how data crosses from one boundary into another. Also, be mindful of the physical infrastructure: have firewalls been updated and have systems been opened to receive and send signals? This seems obvious but it can cost a project days as the project team tries to understand why testing keeps failing.

Data quality

Data is the cornerstone of modern supply chains. Having poor data quality will severely affect the company's ability to be successful, from inaccurate forecasting to how it procures, to how it manages inventory through execution of logistics. A governance framework should be established that the project team can use to follow data standards, policies and procedures. This then helps the project team, who will be validating and cleansing the data so that it is accurate, complete and consistent. The new solution should provide the opportunity to start a new way of managing data going forward, making sure what is inputted into the new system is required and that it meets the standards needed to provide a baseline to create new data from. Providing suitable data training and support to the project team is important and should be run consistently by the company's IT function rather than a one-off exercise to give them an understanding of best practices to follow and ongoing support if issues appear.

EXPERT INSIGHTS

Quality issues with data are present on nearly all technology implementations and are usually in a poor state in the incumbent system or new parameters don't exist that are required for the new system, so data transfers can be redundant in many instances.

You should expect it to be poor and ensure the budget is allocated appropriately here to clean and migrate. To get an idea of what effort is required, during 'business case' preparation, conduct several audits and discuss with suitable experts to be able to cost it. As the project continues, keep the connection between the project team and users. If the users haven't been on the journey, come the 'go live' stage, they may have different expectations of the system's functionality, which can lead to dissatisfaction and them rejecting the system.

Furthermore, many issues with data will not crop up until 'go live'. Testing will identify issues but having a data team mobilized for 'go live' is a must. For example, one retailer at 'go live' had some very odd outputs from its warehouse system, which was instructing a cage fill of flowerpots, but the system was instructing them to be placed side by side on the cage rather than stacked in each other (nest fill), which reduces cage fill of the item (see Figure 12.2). Therefore, the warehouse was trying to ship the dimensions of the cage as though each product needed its full dimensions. This could occur very often across a portfolio of over thousands of different products.

Figure 12.2 Incorrect data and its implications

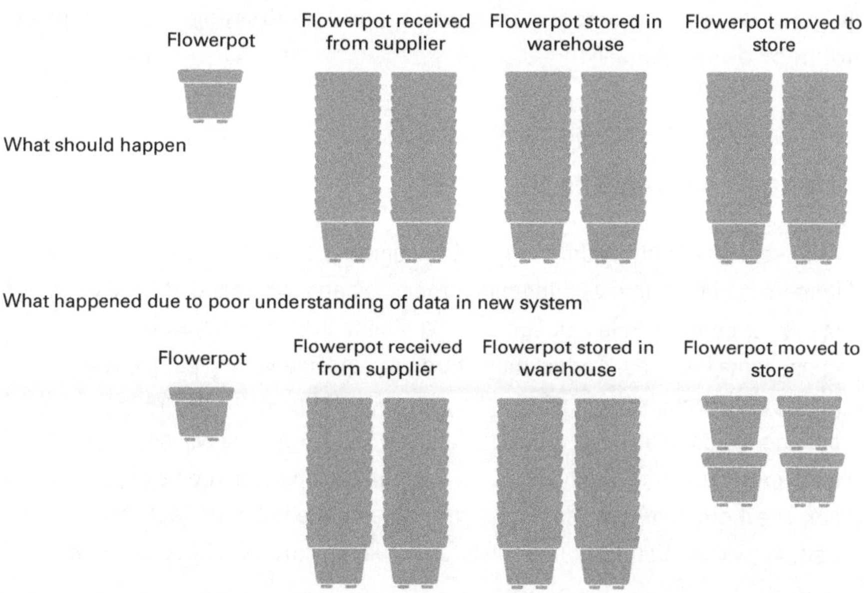

Cybersecurity

When the Wright brothers invented the plane (as they are widely credited with), they also invented the plane crash. Similarly, as technology advances with the use of cloud, so does the 'bad actors' (hackers) ability to infiltrate business systems. Therefore, it is imperative to put in place robust security measures with the ability to audit supply chain solutions to determine how well they are protected, including third parties that engage with the company. Continuous monitoring of this is important and while it is not the priority of supply chain users, they need to be mindful of it and make sure their engagements with third parties adhere to your company's requirements. Furthermore, developing response plans and testing them can help you understand where there are potential risks.

Change management

Managing change effectively starts with clear communication between all relevant parties involved in the project and the business, keeping them informed of progress and using set pieces, such as status meetings, to tackle risks and issues. Having a change management plan developed and actioned will outline the key activities to be completed and should be focused on stakeholders, communications and training requirements. Success or failure following change comes down to how well the users are equipped to use the new solution: proactive monitoring and addressing user feedback should be done regularly.

EXPERT INSIGHTS

At the outset of putting the project team together, the need for experienced leaders is clear. However, updating processes and systems that have been in use for considerable time requires a balance of staff, contractors and consultants involved. Across those three groups, identifying people who have been through change before is crucial, and the staff's desire for change needs to be evident, since then you will know they will go the extra mile for the project, which will count through the ups and downs of project life. The more company resources you can allocate to the project, the better it will be. Contractors are useful; however, care should be taken as they can be a one-person business and therefore the risk is higher (i.e. sickness, better job offers).

Keeping teams motivated through potential long programmes of work comes down to communication; over-communicate and focus on the benefits of the change. Do walkabouts with the senior team and team members and engage in discussion groups with staff on their issues and how the changes will benefit them. Injecting fresh talent into the team is important, as different skillsets are required throughout the life of the project and the expectation that everyone who started the project on day one will be there at day 200 hundred is not right or realistic.

Engagement of stakeholders is crucial. Here, an expert in internal communications who can simplify the key messages for the staff on the changes that are being planned, is very helpful. Engaging the workforce early and unions, if necessary to get their buy-in, is also essential. All the relevant people from different functions that will be impacted by the change should be brought together for a town hall-style briefing and sent regular briefings. The make-up of the project steering group must have decision makers involved from every relevant function affected by the change as a bare minimum. Also, you should attend board meetings, not only for funding but to give regular updates on progress and if further support is needed. One point to note here is the importance of engagement with people in adjacent functions – understanding their perspectives and how the change in supply chain will (or may) affect them is essential.

Getting functional people freed up from day jobs can be tricky even though it may have been agreed many times verbally in advance. When doing 'day in the life' exercises, make sure the focus is on walking in the shoes of the customer, as the technological solution should be about the deliverable of a solution that benefits the customer, the end user.

During the 'go live' phase, if you are not 100 per cent confident, do not allow yourself to be forced into pushing forward. Have tolerance levels agreed on what success / failure equates to and stick to them; plus, have them documented and approved by project sponsors. Executives can look for people to blame when there are major issues rather than focusing on resolution, so make sure to stick to the facts to drive outcomes.

Spend more time on the change aspect and people capabilities rather than the technical elements: people deliver the change and the benefits, not the technology.

Vendor evaluation

Managing different third parties can be time-consuming and costly if not structured. Early on, criteria should be set out for vendor selection and evaluation (I would expect most companies to have such a process in place). Once the vendor of the solution is in place, decisions will be required on business and system integrators, as some companies prefer to take on the roles with these parties, especially within the business integrator space, such as change management and training. Like the vendor of the solution evaluation, understanding the experience they will provide, and the costs is obviously important. Don't be swayed by an analyst's view only; how other clients experienced working with the integrators should provide you with confidence. When negotiating, make sure that a fair comparison can be made and request breakdowns of cost and assumptions. Finally, agree with different parties on how you, as a collective, will communicate, because like putting any team together, there needs to be time to get the team working effectively.

Anticipating where obstacles and pitfalls could come from will help your company proactively plan and execute successful technology projects.

EXPERT INSIGHTS

When going through the process of identifying the right vendor and solution, you need to be comfortable that they will be able to be physically present in the place of work, as many of the salespeople you meet upfront will have little input once the implementation stage arrives. Test their real-life experience: have they worked in supply chain and the sector before? Can you trust them when problems appear?

Do your research and walk the talk, i.e. visit the reference sites, but ask to speak to the people operating the processes. Also ask your vendor to take you to a customer who had difficulties and ask for their lessons learnt.

At the end of this phase, be careful about future 'change control', which can be an ambiguous area in relation to scope and responsibilities. Time should be devoted to the planning and contracting so every party knows what is expected of them. Yes, there might be changes required due to estimations made at this stage, but expectations need to be initially set.

You should also note that IT and procurement teams will also be involved in this process, and it is key to have them all aligned. IT can play an important role: they might be conservative and not support the disruptive technology that can undermine the overall benefits of the new tool, while the procurement team can exclusively look at the cost of running the new tool in licensing fees and shift the decision towards a cheaper, simpler tool that has little to no breakthrough modern technology. It is critical that the four parties (i.e. supply chain, finance, IT and procurement) work together to create the foundation for an unprecedented digital transformation, even if on paper, it looks a bit more risky or financially more challenging, while in other cases, those four players have completely opposite agendas, and coming to a conclusion on the new solution takes a very long time and can be an exhaustive, political process.

Conclusion

There are many facets to successfully implementing supply chain technology and while they may seem on paper relatively straightforward, in my experience, this isn't the case. By understanding the steps, from creating the need for change through the many activities that need to be completed, troubles can appear and how you address them will define the success of the project.

Strategies for overcoming these have been focused specifically on integration challenges, data quality issues, change management issues and third-party partners selection and management. Anticipating issues requires supply chain leaders to keep looking ahead while balancing day-to-day ups and downs. There is not one project I have completed that hasn't suffered from one or more of these issues; however, with a strong project team and sponsorship, issues can be worked through, plans reiterated and lessons learnt to keep pushing forward.

By having a keen awareness of these issues, companies can be better prepared to deliver supply chain technology solutions that will achieve the benefits expected, improve users' experience and strengthen customer satisfaction.

Endnotes

1 Young, L (2024) Target's supply-chain chief turned to stores to answer the online fulfillment challenge, *The Wall Street Journal*, www.wsj.com/articles/targets-supply-chain-chief-turned-to-stores-to-answer-the-online-fulfillment-challenge-167b20bc?utm_source=chatgpt.com (archived at https://perma.cc/FBU8-ALCY)

2 Irani, Z, Abril, R M, Weerakkody, V, Omar, A and Sivarajah, U (2023) The impact of legacy systems on digital transformation in European public administration: Lesson learned from a multi case analysis, *Government Information Quarterly*, 40 (1), www.sciencedirect.com/science/article/pii/S0740624X22001204 (archived at https://perma.cc/Z7FC-68KD)
3 Akella, J, Buckow, H and Rey, S (2009) IT architecture: Cutting costs and complexity, McKinsey & Company, www.mckinsey.com/capabilities/mckinsey-digital/our-insights/it-architecture-cutting-costs-and-complexity (archived at https://perma.cc/MG3F-89JN)
4 Esherwood, P (2020) BP leveraging SAP technology to tackle the world's hardest energy problems, *ERP Today*, https://erp.today/bp-leveraging-technology/ (archived at https://perma.cc/AC7S-8HN3)
5 ASCM (2020) ASCM Supply Chain Operations Reference Model SCOR Digital Standard, ASCM, www.ascm.org/globalassets/ascm_website_assets/docs/intro-and-front-matter-scor-digital-standard.pdf (archived at https://perma.cc/K96E-EPEL)
6 Vegter, D, Van Hillegersberg, J and Olthaar, M (2020) Supply chains in circular business models: processes and performance objectives, *Resources, Conservation and Recycling*, 162, www.sciencedirect.com/science/article/pii/S0921344920303633 (archived at https://perma.cc/RJ9S-W2XC)
7 ASCM (2023) SCOR Digital Standard Quick Reference Guide, ASCM, www.ascm.org/globalassets/documents--files/corporate-transformation/scor-ds-digital-guide_final.pdf (archived at https://perma.cc/GP8T-Q2Z4)
8 Shimange, N (2025) Understanding parent-child associations in data architecture integration, Struto, www.struto.io/blog/understanding-parent-child-associations-in-data-architecture-integration#:~:text=divisions%20and%20brands.-,Parent%2DChild%20Associations%20in%20Technology,compromising%20on%20coherence%20or%20integrity (archived at https://perma.cc/KU5T-5X98)
9 Carr, A (2023) 4 common master data management implementation styles, Stibo Systems Master Data Management, www.stibosystems.com/blog/4-common-master-data-management-implementation-styles#:~:text=A%20Coexistence%20style%20allows%20you,as%20in%20the%20application%20systems (archived at https://perma.cc/A7JF-HXUH)
10 Miseviciute, D (2025) Scrum sprint project management: A detailed agile workflow guide, Teamhood, https://teamhood.com/project-management/sprint-in-project-management (archived at https://perma.cc/Z6Z8-GRHZ)
11 Jones, D (2025) Blue Yonder investigating Clop ransomware threat linked to exploited Cleo CVEs, *Cyber Security Dive*, www.cybersecuritydive.com/news/blue-yonder-ransomware-cleo/737700/#:~:text=Blue%20Yonder%2C%20a%20major%20provider,operations%20at%20those%20respective%20customers (archived at https://perma.cc/NQT6-Y42P)

The future of supply chain management 13

Challenges, trends and opportunities

This chapter will discuss emerging trends, challenges and opportunities in supply chain management.

CHAPTER AIMS

- Stay ahead of challenges, emerging trends and opportunities in supply chain management.
- Appreciate the importance of innovation and growth in staying competitive.

'Life is an ironic tragedy; it has to be lived forward but only makes sense in reverse.'

SØREN KIERKEGAARD[1]

As we become increasingly interconnected across the world, supply chain management must evolve to meet the demands of a changing global landscape, facing the ever-increasing complexity of operating 24/7 with the need for agility, visibility and resilience. This is not only to respond to the challenges of the day but keep the supply chain on a trajectory that foresees future requirements that meet or exceed customers' expectations.

EXPERT INSIGHTS

IBM Infrastructure business has been on a transformational journey to the extent that it is part of the vernacular rather than a phase of work. The manufacturer sells servers, mainframes, storage systems and software with after-sales support. It had revenues of more than $12 billion in 2024, operating in more than 170 countries with assets in 48 countries (e.g. manufacturing sites, warehouses) and over 350,000 SKUs[2] (a mix of finished goods, subassemblies and parts).

It has been on a supply chain transformation since 2010, when focus was on harmonizing it with consistent global ways of working, then moving to a transparent supply chain focus in 2016 and then to an AI-enabled supply chain orchestrated by intelligent workflows from 2019.

In 2025, an intelligent platform that provides a consistent place for users to engage with one integrated and synchronized end-to-end data view was at the centre of its supply chain. This visualization tool removes the need for experts in different applications because they engage at this level using natural language programming (i.e. asking questions) of the system. It also has a transparency layer, which explains the recommendations it is making using AI and frees up its supply chain users to focus on the important issues of the day rather than having to do many manual activities. The transformation involved bringing data together from over 20 applications and it has consolidated to a single instance for ERP and one legacy ERP for the after-sales supply chain.

This is an ongoing venture, but it shows a good example of an approach to continuously evolving a supply chain through transformation. The benefits by the end of 2024 were $200 million cost savings based on the last three years with an improvement in cashflow of $1 billion, accelerating issue resolution from days to seconds, as well attracting a new generation of supply chain talent that wants to work with advanced supply chain capabilities.

Black swan events

Black swan events, as mentioned in Chapter 9, are rare, high-impact events. Take the Covid pandemic, beginning in December 2019, which is a topic that will likely be written about for the next 100 years when discussing the supply chain. It cannot be ignored as it put the role of the supply chain at the forefront of C-suite minds.[3] It highlighted how fragile companies are when black swan events occur, let alone continue and become the norm for which to operate under. There are many examples from this time of where supply chains were pushed to the limits. We will briefly explore three of these below, just to give a flavour of the importance of supply chain in everyday lives.

Starting with a product many people take for granted: toilet roll. For context, approximately 83 million[4] rolls are produced daily using 0.4 rolls on average per day.[5] The raw materials (paper fibre and water)[6] are available in abundance and demand is stable. It was deemed that manufacturers had sufficient capacity to cope with typical demand, so then why did we have a crisis?[7] We had one because consumers reading the news and talking to family and friends became frightened and panic-bought months' worth of purchases in one day. This caused a spike in demand whereby manufacturers could not increase capacity quick enough because this was normally a stable product with consistent demand and therefore operations had been set up to treat it this way. This led to the 'bullwhip effect', which amplifies a shock from consumers back through the end-to-end supply chain.[8] Retailers reacted by trying to prevent consumers purchasing more than one box or bag of 12–18 rolls, however by the time they had enforced this, it was too late. It took government intervention and manufacturers time to get back to restore a stable supply again. In classic economic terms, we even saw the effect of a product move from having price inelasticity to elasticity, with marketplace sellers offering it at high prices.[9]

Humanitarian supply chains, as another example,[10] were terribly affected during the pandemic. Governments mandated closures of borders and shops. Enforced social distancing and staff shortages led to an unprecedented demand for personal protective equipment (PPE) products and in the race to secure enough supplies, at some points it became a point of national pride to secure inventory.[11] This put countries stricken with wars and poverty versus wealthy nations, who were all looking for the same products and led to a world order based on those who could pay, gaining and then facilitating the supply chain to get it to their markets. The products were further restricted by transportation movement and there were delays due to reductions in commercial flights and tighter point of entry controls. This also led to price instability and, following classical economic theory, drove higher inflation that took many years to self-correct.[12]

A final example, which continues to be a challenge but was amplified during Covid, was in relation to the semiconductor chip market, which faced difficulty on numerous fronts from shifts in demand, supply shortages and geopolitical turmoil. From a demand perspective, the shift to working from home created a surge in demand for laptops and smart devices and a reduction in consumer electronics (e.g. phones) and vehicles. This exposed the vulnerability of chips. Supply shortages came from closures of manufacturing plants and the inability to move chips across the globe. Also, the type of chip to make was a large consideration, since the process for making them can be very long (e.g. up to 1,400 process steps depending on the complexity of the chip, with manufacturing cycle times of up to 26 weeks).[13] More than 70 per cent of chip manufacturing is executed by companies including TSMC (Taiwan) and Samsung (South Korea) in East Asia.[14]

These examples and many others at this time, show that our reliable supply chains can quickly be pushed beyond their comfort zones. Many companies that up to then would have stated they had resilience to cope were proved wrong.

Geopolitical trends

Unrest, such as the Ukraine War (starting in February 2022) and the Gaza War (starting in October 2023), has significant implications on the supply chain. It is worth noting that wars create their own supply chains, but here I will focus on global supply chains and trade. Trade routes are usually the first to be choked by conflicts, impacted by port closures and damage to transport links, for example, which can reduce the movement of products or mean that alternative routes from source to the customer must be found. This leads to higher transportation costs and longer lead times. In other instances, if the country at war is a provider of key raw materials (e.g. Ukraine and grain), then shortages of inventory will happen. Companies then must find alternative countries to source from, which can lead to fluctuations in pricing and very different costs to serve. The Houthi attacks near the Suez Canal, which started in late 2023, led to a 66 per cent decrease in traffic on that route.[15] The price of a container went from just under $1,500 to over $5,000 and in March 2025, was over $3,000.[16] Also, the alternatives such as re-routing via the bottom of Africa are estimated to increase cost per vessel of over $1million,[17] so considering that 12–15 per cent of global trade goes via the Suez Canal,[18] the implications are significant. Therefore, the price of sending items via this route, as well as the risk of attack, led to higher supply chain costs.

Also, in March 2025, global supply chains faced a 'tariff' war where the demand for 'local' versus 'global' supply chains were being tested. The US government agenda was to encourage the movement of manufacturing to the US and where this was not possible, it stated that countries could expect to see additional costs incurred to sell to US consumers. Making changes to global supply chains presents multiple challenges, including the requirement to set up new routes for supply and to service demand, let alone manufacturing taking considerable time and investment. For example, setting up a new manufacturing site and have it running properly will take more than a year, if not multiple years, depending on how advanced its requirements are.[19] By the time the site is operational, policies may have moved on and the cost to do business could make the site redundant before it has even started to make products. There are many reasons beyond this book why operations are done where they are (such as chasing the lowest labour costs), but whatever the reason, the point is that supply chains can't pivot as quickly as government policies. Take the action by the Liquor Control Board of Ontario, which purchased over $1 billion of US alcohol.[20] In response to tariffs put on it from the US government, at the beginning of March 2025, the state removed from sale US-made drinks, which had significant ramifications on the supply chain. Firstly, there was a problem of what to do with the inventory at store and bar level. Sending it all back to a consolidation point and then shipping it back to US suppliers would be extreme, since this was a tit-for-tat confrontation. Bringing inventory to a consolidation point and then refunding the owners would be one strategy, but it would then impact inventory levels, which would need to be reset, with production delayed, avoiding overstocking. Companies in these types of situations need to understand and decide whether they believe this is a short-term decision which will be reversed quickly or whether they need to ship products back to other markets, and what to do with demand and manufacturing. This explains why supply chain decisions need to work fluidly and can't be too reliant on predictable behaviour.

Consumer-led and company trends

Diversification and agility are important for supply chains and are therefore becoming a trend for companies. Companies need to have predictability and stability in how they operate, which can at times be in conflict to lowest cost. They need to understand where they have risk, whether that be at a macro level (e.g. political, economic, social) or supplier related. To deal with macro level risks, companies having operations in multiple locations and the ability to move manufacturing across geographies in case something happens in one location or the cost to serve changes materially, is advantageous. Nearshoring (i.e. setting up operations in countries / regions close to market serving) and localization (i.e. making and selling in the same country) evolve to mitigate risks from some of the themes already mentioned. Also, companies need to understand inherently that the supply chains that support your supply chain in both directions are important and need to be managed proactively. Diversification (i.e. not reliant on one supplier or location) makes sense and being able to have multiple suppliers and sources of products helps in negating risk. It also assists in balancing out risk on the country of origin and being ready to make changes based on a variety of factors that can be from weather to the price of doing business in other countries. However, challenges need to be evaluated and while procurement will rightly be focused heavily on the price point, having only sole supply in most cases will increase the risk significantly. Therefore, trade-offs will need to be discussed and agreed upon between different functions. For example, manufacturing of a specific product might only be able to be completed in one location due to a multitude of reasons; therefore, the ability to switch on alternative supply may not be available to the company (think of the semiconductor chips example earlier). In those instances, knowing what to do is crucial. Many times, the customer will only become aware of a problem when there is a failure of supply and by this point, trying to tackle it could take considerable time to solve.

The trend of traceability, which we have covered in previous chapters, is only going to become more important and a key consumer consideration for many purchases now and in the years to come. Having visibility of the different stages of the supply chain from source to customer and the players who participate throughout is going to have to move beyond basic set-ups, which are typically visibility of your first-tier suppliers and their locations only.

Traceability can be seen as an enabler for social responsibility and sustainability as well. For example, slavery has still not been eradicated from global supply chains. Global regulation from the United Nations has been driving the need for it to be completely removed for nearly 100 years[21] and there has been rightful outrage when it has been discovered that companies might be neglecting their duty of care, even unknowingly, like in relation to labour in distant parts of their supply chain. Take the cocoa farms in West Africa that were exposed by the BBC[22] in 2019 for using child labour. The case, when brought to the public's attention, was tackled but this was by no means the only example. Keeping control of such challenges is not easy when operating globally. A hindrance to tackling this challenge is the lack of consistent standards and regulation between different countries, which makes it harder for companies to know and follow the rules of business and to operate with transparent supply chains. Ignorant companies can become very unpopular with consumers when these issues are highlighted and can result in significant reputational and financial damage. Therefore, the need for companies to know their supply chain end-to-end and ensure all players who participate in it are compliant to global laws, regulations and standards is crucial.

From a sustainability perspective, the supply chain has an increasing need to explain greenhouse gas emissions and other unsustainable practices that can occur through their upstream supply chain, in their operations and then downstream to the eventual customer. Companies are gaining an understanding of their impact and then determining how to keep track of it and improve with the use of technology.[23] For example, the United Nations found in its research that almost a third of all food is wasted.[24] In 2022, the UK grocery market started eliminating expiration dates on perishable goods to tackle food waste.[25] The results so far have been mixed because there are still fundamental education gaps in what consumers understand by different terms, which can appear on products such as 'display until' and 'use by'. This is linked to why 10 per cent of Europe's food is wasted.[26] Also smart packaging is being tested and deployed as a solution to improve the integrity of a product through its journey, as well as using technology to enable the consumer to know what they have in the fridge, what has gone off or nearly run out, and then automatically send a signal to replenish to their local retailer, for example, or create a list on their phone to inform them what to buy in the grocery store.[27] Also, there have been changes to EU law in relation to charging sockets, which again was an approach to tackle waste. In October 2022, the EU's Common Charger Directive was approved, making it mandatory that all devices support USB-C charging. This combats how many households have countless different sockets (at a cost estimated to be €250 million to EU consumers)[28] and significant e-waste,[29] which is estimated to be 11,000 tons a year.

GOBI CASHMERE AND AGI DENIM

Gobi Cashmere and AGI Denim are examples of clothing companies that are addressing traceability, social responsibility and sustainability. As well as this, they are showing foresight about future legislation and are acting now to be differentiated from their competitors, serving customers directly overseas or supplying to manufacturers further down the supply chain.

Gobi Cashmere sources its raw materials from various provinces in Mongolia, which allows it to maintain complete control over every stage of its manufacturing process. This vertical integration ensures high-quality and sustainable production, working closely with Mongolian herders to ensure ethical and responsible sourcing, aligning with the nomadic culture's emphasis on the relationship between people and the environment.

AGI Denim, based in Karachi, Pakistan, sources around 50 per cent of its cotton locally from the Sindh and Punjab provinces, with the rest coming from various international locations. This localized sourcing reduces lead times and transport costs, while also lowering the carbon footprint. It has developed systems for organic, regenerative and recycled cotton, and implemented measures to reduce carbon intensity. The company's vertical integration allows it to control the entire production process from the cotton farm to the final garment, enhancing efficiency and traceability.[30] We looked at that in Chapter 8 – the three aligned trends of traceability, social responsibility and sustainability.

Enabling customization and personalization are trends that are disrupting supply chains as well as enabling company differentiation. Many companies speak of a customer-centric model, however, putting it into practice needs a level of flexibility such as delaying final manufacturing of finished goods, allowing customers the opportunity to alter specifications up until the last possible moment and leveraging analytics to better understand what customers will want based on a multitude of variables (e.g. sales, season, what celebrities are wearing, weather, interest rates etc).

Ralph Lauren provides the option for customers to add their initials to their clothing and Net-a-Porter offers value-added services such as personal shopping support. These companies see customization as a way to keep customers loyal. Net-a-Porter has stated that 2 per cent of its customers drive 40 per cent of sales.[31]

Flexible production has also led to advancements in last-mile delivery innovations. For example, Amazon's ability to orchestrate deliveries within an hour in some cities sees next-day as its standard way of operating. This has led to customers embracing e-commerce and the bar has been set for other companies to match. Innovations for the supply chain which have come from this are primarily in the logistics arena and include drones, autonomous vehicles and delivery platforms (e.g. Uber, Deliveroo), which we have discussed previously, where companies meet customer demands for speed and convenience.

Digitalization trends

The ever-evolving technology landscape is not waiting for companies to get comfortable with many of the digital capabilities available today to aid in the running of global supply chains. Many of the examples discussed in the following section will continue to challenge decision makers on the steps they need to take to keep their companies moving forward, competing and winning in their markets.

We will look at a few of these digitalization trends now and determine where they could be applicable in supply chain management.

Generative AI

From 2023 onwards, there was much hype focused on Generative AI (GenAI) and all the amazing things it can potentially offer companies. GenAI is a form of machine learning (a subset of AI) that can generate new content or data of many varieties (i.e. audio, video, images, text, code, simulations and 3D objects) from asking it questions through prompts. It is advancing the capabilities of traditional AI and analytics, augmenting them in areas such as document summarization, conversational search, content creation and code generation. However, the adoption of GenAI does come with risk and must be managed carefully, such as in the role of ethics, bias, design standards, prompting guardrails, secure tools and quality standards.[32]

It is a priority for the C-suite (maybe due to fear of missing out) as the promise of productivity and efficiency gains cannot be ignored.[33] It most certainly cannot be argued that it is following 'Amara's Law', where the impact of new technology is overestimated in the short term while its long-term effect is underestimated,[34] like blockchain. However, from a company perspective, its true potential can be unlocked once you start building your own language models using your own company data, instead of outdated or incorrect information from the internet.

In relation to supply chain management, there are several perceived benefits that it could potentially capture and you won't be surprised that a few of these keep coming up, which we will discuss in due course. Demand forecasting was always going to be an area of focus for GenAI. Forecasting visibility can come from being able to use GenAI to digest various forms of data from the company's systems, translate them into a standard form and then use prompting to find out what is recommended. For example, knowing the effect a hot day in London has on ice cream sales will help to direct inventory to well-performing locations there to minimize lost sales. Furthermore, knowing when hurricane season is coming will drive demand for nail guns as people board up their windows; post-storm demand goes up for equipment to demolish buildings that were hit.

GenAI is risky if it is not explainable, due to its 'black box' nature, where the user cannot see how it comes up with an answer but in time, this uncertainty will reduce as planners gain confidence in the recommendations.

Where traditional supply chain management often involves reactive strategies, where leaders must constantly adapt to unforeseen challenges, GenAI could significantly enhance the ability to pre-empt supply chain disruptions. By leveraging machine learning, automation and advanced analytics within a hybrid cloud environment, companies can anticipate demand fluctuations, sourcing delays and other potential disruptions and be proactive about these.

Moreover, GenAI could transform workflows within supply chains. According to research by IBM, 64 per cent of chief supply chain officers acknowledge that GenAI is completely transforming their workflows.[35] This transformation is largely due to the automation of traditional and transactional processes. In time, it could lead to AI assistants handling most of these processes, allowing human workers to focus on more strategic tasks. This shift could not only increase efficiency but also enhance the overall agility of the supply chain.

GenAI could also play a crucial role in driving sustainable innovation. By analyzing data from various sources, it can identify opportunities for reducing waste, optimizing resource use, and improving overall sustainability. This could not only benefit the environment but also enhance the company's reputation and compliance with regulatory standards.

GenAI can also be envisioned as a translator – putting data from the many different systems you use in the supply chain into one consistent data set that then feeds a single user interface.

Overall, GenAI could become essential for modern supply chains due to its ability to predict disruptions, automate workflows, enhance decision-making, drive sustainable innovation and improve visibility and collaboration. By harnessing the power of GenAI, companies could transform their supply chain strategies, achieve greater efficiency and maintain a competitive edge in an ever-changing business landscape.

Agentic AI

Agentic AI refers to autonomous systems that can make decisions and take actions independently to achieve specific goals. Unlike traditional AI systems, often referred to as passive AI, which follow predefined instructions, agentic AI operates with a degree of self-direction, like having a smart assistant that can proactively handle tasks and solve problems without constant supervision.[36]

The concept of agentic AI represents a significant evolution in AI technology. Its level of autonomy is made possible through advanced machine learning algorithms, which enable the AI to learn from its experiences and improve its performance over time. The development and deployment of agentic AI also raise important ethical and regulatory considerations. Ensuring that these systems operate transparently, fairly and safely is crucial to gaining trust and maximizing their benefits.

Examples of use within supply chain management are varied and can cover many aspects. At a general level, agentic AI can automate routine and repetitive tasks, freeing up users for more strategic and creative work. This can lead to significant productivity gains and improved efficiency. It can improve decision-making through being able to analyze vast amounts of data, identify patterns and make decisions in near real time, often faster and more accurately than human users. This can also lead to reduced errors. It can drive improvements in customer experience by enabling personalized and proactive customer service, improving customer satisfaction and loyalty through chatbots and virtual assistants that are always available to answer frequently asked questions and helping customers with simple issues. Agentic AI can provide detailed records of all interactions, decisions and actions, enabling transparency and accountability. It can monitor and respond to safety and security threats in near real time, reducing the risk of accidents and cyber-attacks. It can also assist with autonomous vehicles in the warehouse and delivering, for example, in detecting and responding to hazards more quickly than human drivers.[37]

McKinsey[38] in 2016 predicted that through digitalization that 80–90 per cent of all planning tasks could be automated. Planning systems and process automation have aided planners but not to the level predicted by McKinsey, with suggestions that only 39 per cent of planners are using the new tools provided.[39] However, now with the emergence of agentic AI, there is a further opportunity to eliminate many time-consuming activities for a supply chain planner.

For example, it isn't a stretch to imagine a planner having various digital agents assisting them with their activity list, such as tasking one agent with identifying what the priorities are for the day based on scanning underlying systems data and then feeding that to another agent to determine which can be actioned. Another agent could write updates back to the systems without disturbing the planner and only provide the planner with the exceptions that need their expertise and critical thinking.

Agents could also be used to help the planner by:

- cleansing data in a particular category
- reviewing data from various perspectives (e.g. historical, external news, weather, local event schedule) and putting forward recommendations for any change to the forecasts for review
- doing a deep dive into specific categories understanding why any forecasts were missed
- reviewing demand and testing multiple scenarios based on input from a production agent
- assisting the planner prepare their inputs for various weekly and monthly cycles such as the S&OP process.

Figure 13.1 shows an example of how digital agents could operate in the hands of the planner.

For procurement, agentic AI could be deployed to help a user determine what to do in relation to supply. One agent can evaluate all the suppliers of a raw material from a performance view (e.g. quality, on time, price), another can look at when these materials are needed for production and another can review tracking sales in near real time, which provides the user with information to make the right decision or the system can make the decision, providing reasoning for why that decision was made.

From a logistics perspective, the agentic AI could autonomously manage different parts of the operation, with one agent looking at routing in a specific area, trying to optimize for delivery times and costs, while another focuses on sustainability impacts of potential decisions, while another is

Figure 13.1 Planner engaging with digital agents

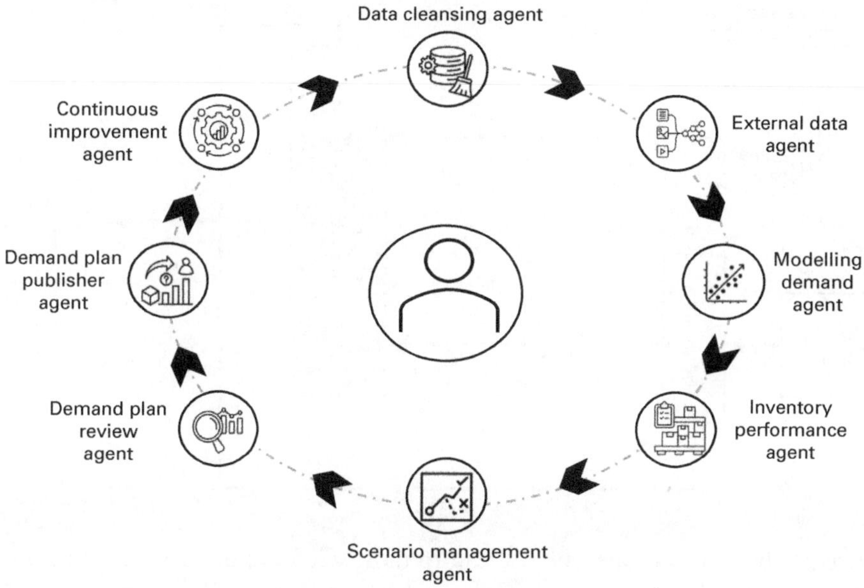

monitoring traffic and weather patterns. These working independently feed their views into another agent, which the user is using to consolidate inputs to make decisions based on trade-offs all in near real time.

Generative adversarial networks

Generative adversarial networks (GANs) are an approach to modelling using deep learning methods. Primarily known for their applications in image generation, at a conceptual level, it is a type of AI that consists of two neural networks (a generator and a discriminator) that compete against each other, enabling the system to generate new, previously unseen data based on the existing dataset. Unlike digesting unstructured variable datasets as described in Chapter 5, here it is treated as deep learning. While the concept of GANs was developed by Ian Goodfellow in 2014, it is still an emerging technology in relation to supply chain and examples of its use are still primarily immature.[40] However, they can potentially be applied to supply chain management in a variety of ways.

In relation to demand forecasting as seen in Figure 13.2, GANs can be used to generate data that resembles historical sales data (which can help minimize distortion from exogenous shocks such as Covid or the Suez Canal) and allow companies to train their demand forecasting models more

Figure 13.2 GANs for demand forecasting

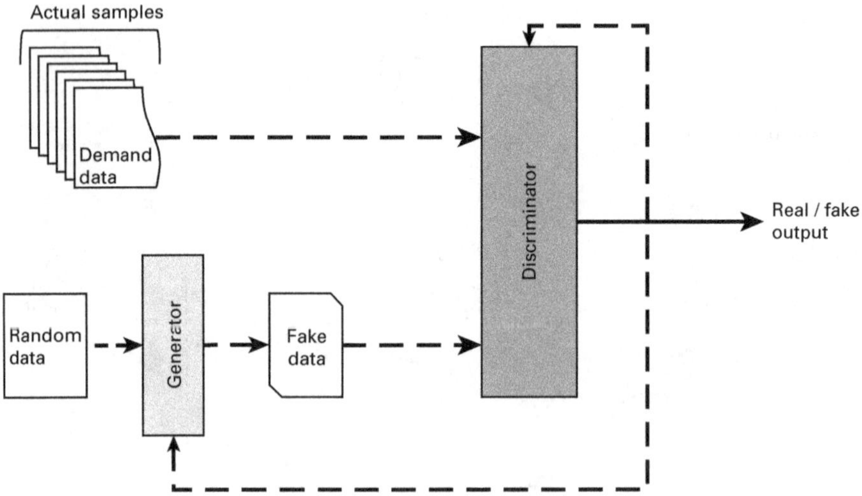

effectively. This, in time, should lead to more accurate predictions of future demand resulting in benefits such as improved inventory management and minimizing stockouts or overstocks. GANs could be used to generate synthetic data that could represent different supply chain scenarios, such as disruptions. By simulating scenarios in a 'safe' environment, companies can understand their resilience and develop rigorous risk mitigation strategies. Optimization of schedules such as production schedules could be explored by GANs generating optimal scenarios based on historical data and constraints, assisting companies to maximize production efficiency, utilization, reduce waste and minimize costs.

GANs could be deployed to evaluate supplier performance (e.g. lead times, quality and reliability) by feeding data to training models, which rank them, enabling companies to have more control and insights to inform decisions on supplier selection. GANs are already being deployed in relation to predictive maintenance, which helps companies simulate equipment failure patterns and generate scenarios of wear and tear to proactively schedule maintenance activities and reduce downtime. In time, as the technology is piloted more and its use becomes more popular, GANs have the potential to help companies transform their supply chains.

Quantum computing

Quantum computing is an exciting field[41] which again has the potential to revolutionize how companies run their supply chains. It has capabilities beyond classical computing (which relies on bits as its smallest unit of data, represented by either 0 or 1), whereby quantum uses quantum bits (qubits). Qubits can exist in multiple states simultaneously following a phenomenon known as 'superposition'. What this means is that quantum computers can process a vast amount of information at once, offering users the ability to do computations rapidly.[42]

One of the key principles of quantum computing is 'entanglement' (bear with me as I know this is getting technical!) which is where qubits interconnect such that one qubit directly influences the state of another regardless of the distance between them. In simple terms, following this process enables quantum computers to perform complex calculations more efficiently than classical computers. It is argued by those leading the research in this area that this is not an expansion of classical computing since it requires a different approach to problem-solving. Problems need to be decomposed, executed in parallel then reconstructed. There has already been concerns raised about its computational power and its ability to break encryption schemes. Therefore, as this field develops, a quantum-safe approach is required to secure data against cyber-attacks.[43]

When we look at quantum computing from a supply chain management perspective, it is not far-fetched to see how it could help with solving complex optimization problems much faster and efficiently than classical computing.

Looking at Figure 13.3, the obvious first stop for potential opportunity is in the optimization of routing for companies where quantum computing could determine optimal routes for getting products from A to B. These problems today are solved typically as one-off exercises (unless your organization is a logistics provider that needs to be evaluating this as part of its day-to-day

Figure 13.3 Route optimization with quantum

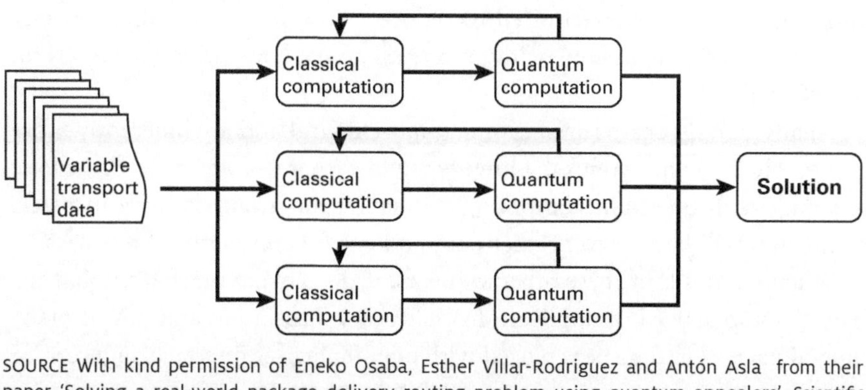

SOURCE With kind permission of Eneko Osaba, Esther Villar-Rodriguez and Antón Asla from their paper 'Solving a real-world package delivery routing problem using quantum annealers', *Scientific Reports*, 14(1), 2024, doi:https://doi.org/10.1038/s41598-024-75572-1

work) and are highly complex, need lots of computational power and are very slow to run. By deploying quantum processing, companies will be able to find optimal routes much quicker, be able to create risk scenarios and even respond in near real time to adjust routes, leading to saving time, reducing fuel consumption and minimizing transportation costs.

Moving on from optimizing routing, the supply chain network design will be the next obvious opportunity for quantum computing. This involves determining optimal locations for assets such as manufacturing plants, warehouses and distribution centres, considering trade-offs of transportation costs, time, demand and capacity constraints. Having completed many network reviews, the pain of running multiple scenarios and having to wait an hour or two to see output, then realizing you forgot to adjust a variable, will soon be a distant memory. Similarly, quantum computing will enable optimizing solutions faster, leading to more efficient and cost-effective supply chain networks.

Demand forecasting will be able to handle even more datasets and not just generate more accurate demand forecasts but also be able to explain each variable and either accept or reject those, which helps to inform a better forecast quickly. This will enable companies to optimize inventory and improve customer service.

Quantum computing, as it will be able to look at global supply chains, assess risks and analyze complex simulations quickly, will improve resilience. The development of sustainable supply chains could also benefit from quantum computing since it will be able to analyze many more datasets than current classical computing and provide different outcomes based on the factors most important to companies, assessing the key trade-offs of cost, service, capital, sustainability and resilience (linking back to the supply chain pentagon we looked at in Chapter 3 and shown again in Figure 13.4).

This capability could lead to vastly improving end-to-end supply chain decisions. For example, through a sustainability lens, being able to compute and optimize whenever required to reduce transport use, minimize energy consumption and emissions with an overall goal of operating a low-carbon supply chain.

Finally, supply chains are becoming more digital and are inherently more vulnerable to cyber-attacks. However, while there is an issue of cryptic breaking with quantum computing, by having quantum security in place, companies will be able to protect data and ensure continuity of business.

When we think of these advancements, it does make me believe that the typical smorgasbord of applications that users must go in and out of many times daily will in time be replaced with one single user interface. Figure 13.5

Figure 13.4 Supply chain pentagon of trade-offs

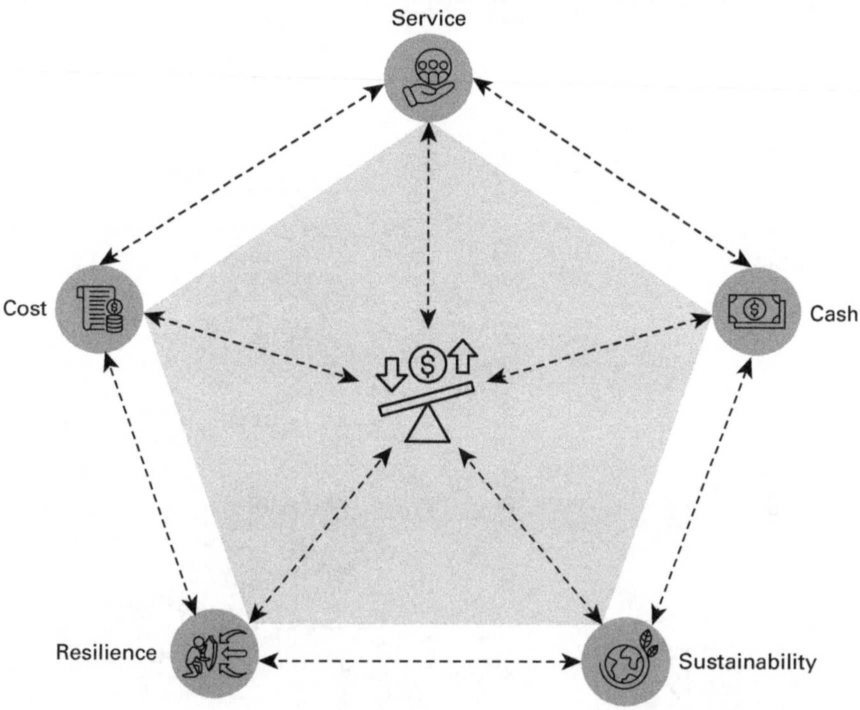

is how I could imagine the supply chain operating at a conceptual level with many of the systems we have discussed in this book, having the ability to collate data, provide it to users in whatever role they have, collaborate with others knowing all are looking at a 'single source' of truth to inform decisions and trigger digital agents to aid them in their jobs.

I do recognize that could be a step too far for many readers today and that utopia for many companies is still a long way off; however, having a vision like this with the steps outlined in Chapter 2 will address many of the shortcomings that supply chain technology has. Although it will be a challenge to receive data at a speed and level of quality you require from external partners, companies getting their own part of the supply chain in order is still advantageous. Furthermore, I can see a future where we don't need to be experts in technology systems: we will use information provided in near real time from them (or the underlying code of the software) and the user interface will enable us to engage with a more advanced form of natural language processing to receive the information we need, overlay our critical thinking in collaboration with others in the supply chain and make informed decisions.

Figure 13.5 Supply chain technology future

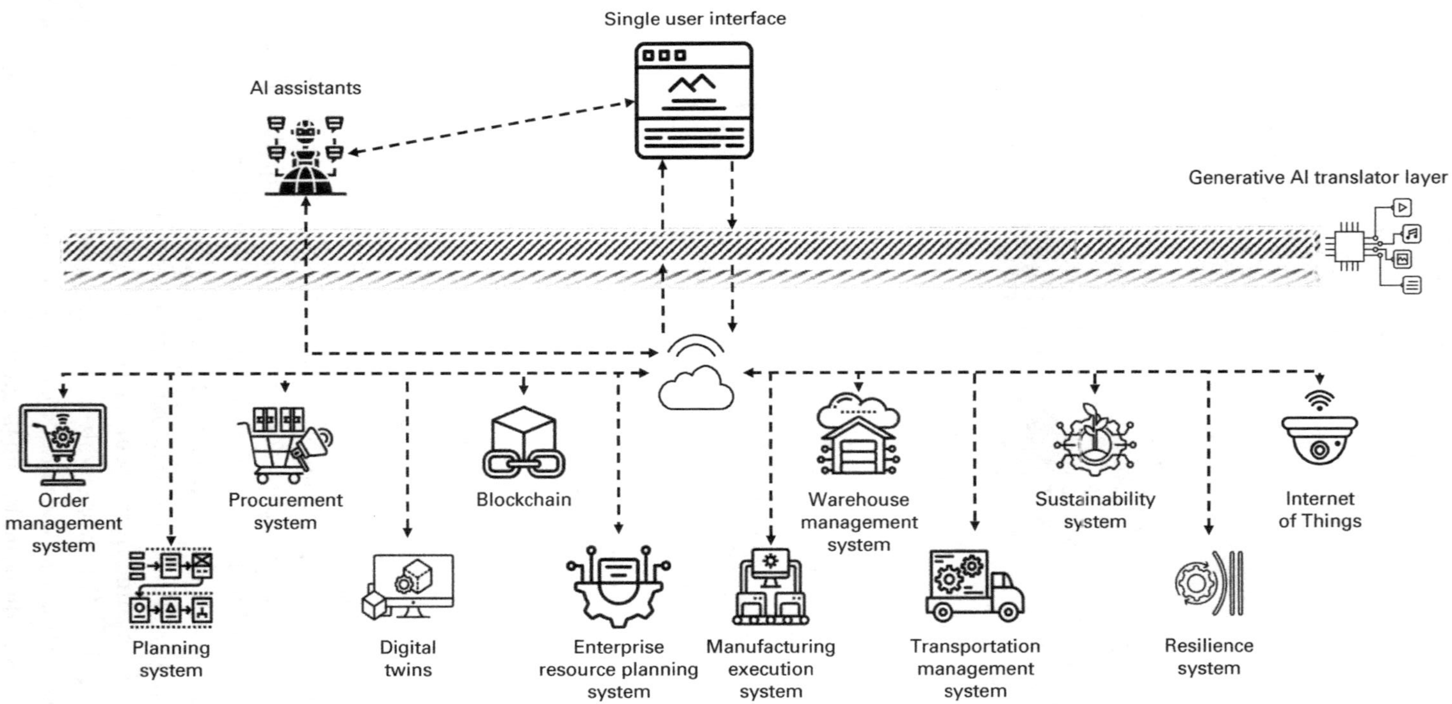

Accelerating the innovation cycle

Consumer-led and company trends and digitalization capabilities are key considerations for companies when it comes to decisions on innovation and growth for their supply chains. Staying competitive requires leaders to focus on the benefits that will come from investing in and evolving their supply chains according to these trends and digital capabilities. All decisions should start with the customer in mind and address their needs and preferences – these don't remain constant for too long, so supply chains need to be designed to adapt and innovate.

Embracing an innovation mindset in the supply chain enables the adoption and acceleration of new products and services, which improve customer experience and cater for a variety of preferences. Adapting to market dynamics, such as black swan events and geopolitical trends already outlined, can significantly impact supply chains, so having a mindset of innovation and growth will aid companies to adapt quickly, be more resilient and better equipped to handle disruption and maintain a competitive edge.

EXPERT INSIGHTS

Bringing supply chain innovation into large organizations needs to be a deliberate activity encouraged by leadership. For example, in some global organizations, during monthly steering committee meetings, time is allocated to bring in new technology and look at it. It is about challenging teams to be curious and get them out of their mindset of execution mode.

Importantly, innovations in the supply chain contribute to environmental sustainability by reducing waste, optimizing transportation routes and promoting the use of renewable resources. Companies that prioritize sustainability are more likely to attract eco-conscious customers and stay competitive in the long run.[44] A culture of innovation and growth also attracts top talent who are interested in working for forward-thinking companies.

By sharing data and co-developing solutions with key partners (e.g. suppliers and logistics providers) in the end-to-end supply chain, companies can create a more innovative, adaptive and collaborative supply chain, and build a skilled workforce that drives continuous improvement that helps to maintain competitiveness. Companies that prioritize innovation and growth in their supply chain strategies are more likely to experience long-term success.[45]

Below are some suggestions to aid innovation and growth when managing the supply chain.

- **Continuous learning and research** – stay informed about the latest industry trends, technologies and best practices by regularly reading industry reports, attending conferences and participating in webinars and workshops. You should also encourage employees to engage in continuous learning and provide opportunities for professional development.
- **Collaboration and networking** – build relationships across your wider supply chain ecosystem with suppliers (at a variety of tiers and not only procurement professionals), customers (engage in quarterly reviews, not just commercial updates but include supply chain elements and representatives), logistics providers and other stakeholders. By doing this, you are enabling collaboration and information sharing, helping to discuss and identify emerging trends and challenges, learning from each other.
- **Sustainability and social responsibility** – incorporate sustainability and social responsibility into the core of your supply chain strategy. Focus on understanding and reducing waste, optimizing transportation routes and where possible, using renewable resources aligned to customer desires and regulation. As well as this, align any identified future opportunities with sustainability and social responsibility initiatives.
- **Agility and adaptability** – design supply chain processes and infrastructure that are agile and adaptable to change through challenging flexible production facilities and diversifying supplier networks. Also, adopt demand-driven supply chain models that can respond quickly.
- **Innovation and experimentation** – encourage a culture of innovation and experimentation within the supply chain by testing new technologies and reimagining processes.

Conclusion

The supply chain management landscape is undergoing significant changes. The increasing complexity of operating in a global economy, combined with the rise of emerging trends and technologies, presents both opportunities and challenges for companies.

The Covid pandemic highlighted the fragility of supply chains and the need for resilience, adaptability and agility. Geopolitical trends, such as the conflict in Europe and the Middle East, also have a significant impact on global trade and supply chains. To stay ahead, companies must adopt a proactive and agile approach to supply chain management. This includes embracing digitalization, investing in advanced technologies such as artificial intelligence and prioritizing sustainability and social responsibility.

The trends and digitalization capabilities discussed in this chapter, including generative AI, agentic AI, generative adversarial networks and quantum computing, have the potential to transform the way supply chains are managed and optimized. However, their adoption will require companies to be 'match fit' in relation to data and processes, and to embrace an innovation mindset, encouraging a culture of experimentation, which prioritizes sustainability and social responsibility.

Ultimately, the future of supply chain management will require companies to be adaptable, resilient and agile in the face of changing market conditions and emerging trends and technologies. By prioritizing innovation, sustainability and social responsibility, companies can build a skilled workforce, drive continuous improvement and maintain a competitive edge in the global market.

Endnotes

1 Kierkegaard, S (2005) *Fear and Trembling*, Penguin Books

2 IBM (2025) IBM 2024 Annual Report, IBM, www.ibm.com/downloads/documents/us-en/1227c12d3a38b173 (archived at https://perma.cc/D55C-5QRU)

3 Harapko, S (2023) How COVID-19 impacted supply chains and what comes next, EY, www.ey.com/en_gl/insights/supply-chain/how-covid-19-impacted-supply-chains-and-what-comes-next (archived at https://perma.cc/MTT9-Z7J7)

4 Hollingsworth, J (2023) Choosing eco-friendly toilet roll for your business, PHS Direct, www.phsdirect.co.uk/blogs/choosing-eco-friendly-toilet-roll-for-your-business (archived at https://perma.cc/XZM5-6JR4)

5 Redd, N (2025) The Best Toilet Paper, *The New York Times*, www.nytimes.com/wirecutter/reviews/the-best-toilet-paper (archived at https://perma.cc/3HY6-Q4Q2)

6 Schrotenboer, B (2020) Coronavirus and shopping for supplies: Getting to the bottom of the toilet paper shortage, *USA Today*, https://eu.usatoday.com/story/money/2020/04/08/coronavirus-shortage-where-has-all-the-toilet-paper-gone/2964143001 (archived at https://perma.cc/W46X-PUWL)

7 Cecere, L (2021) The lessons from the lonely roll of toilet paper in the pandemic, *Forbes*, www.forbes.com/sites/loracecere/2021/04/21/the-lessons-from-the-lonely-roll-of-toilet-paper-in-the-pandemic (archived at https://perma.cc/EA38-92DY)

8 Lee, H L, Padmanabhan, V and Whang, S (1997) The bullwhip effect in supply chains, *MIT Sloan Management Review*, https://sloanreview.mit.edu/article/the-bullwhip-effect-in-supply-chains (archived at https://perma.cc/5LD7-K9YX)

9 Moore, A (2020) How the coronavirus created a toilet paper shortage, College of Natural Resources News, https://cnr.ncsu.edu/news/2020/05/coronavirus-toilet-paper-shortage (archived at https://perma.cc/H5YB-VW8P)

10 Travers, C L and Lowe, A S (2022) Humanitarian supply chains during COVID-19: Systems failures, recovery and emerging alternatives, Royal Academy of Engineering, https://engineeringx.raeng.org.uk/media/sujbzy0j/engineering-x-humanitarian-supply-chains-during-covid-system-failures-recovery-alternatives.pdf (archived at https://perma.cc/T8NZ-ZWMC)

11 Davies, G (2020) The supply of personal protective equipment (PPE) during the COVID-19 pandemic, National Audit Office, Department of Health & Social Care, www.nao.org.uk/wp-content/uploads/2020/11/The-supply-of-personal-protective-equipment-PPE-during-the-COVID-19-pandemic.pdf (archived at https://perma.cc/G9XA-NMFU)

12 Brooks, R, Orszag, P R and Murdock III, W E (2024) The lagged effects of COVID-19 supply chain disruptions on inflation, Brookings, www.brookings.edu/articles/the-lagged-effects-of-covid-19-supply-chain-disruptions-on-inflation (archived at https://perma.cc/JAU4-2G27)

13 Mohammad, W, Elomri, A and Kerbache, L (2022) The global semiconductor chip shortage: Causes, implications, and potential remedies, *IFAC-PapersOnLine*, 55 (10), www.sciencedirect.com/science/article/pii/S2405896322017293 (archived at https://perma.cc/728P-X9KL)

14 Marjanovic, M (2024) 10 biggest semiconductor companies in the world, Finbold, https://finbold.com/guide/10-biggest-semiconductor-companies-in-the-world/#:~:text=The%20top%2010%20biggest%20semiconductor%20companies%20are%20NVIDIA%2C,company%3F%20The%20fastest%20growing%20semiconductor%20company%20is%20NVIDIA (archived at https://perma.cc/4ZVU-E5T4)

15 FDD (2024) Maersk reports Houthi attacks cause 66 percent drop in Suez Canal traffic, FDD, www.fdd.org/analysis/2024/09/06/maersk-reports-houthi-attacks-cause-66-percent-drop-in-suez-canal-traffic (archived at https://perma.cc/98EP-5Z95)

16 Kumar, S S and Sarmmah, S (2024) Shipping container cost guide for smart 2025 purchases, Container X-Change, www.container-xchange.com/blog/shipping-container-price (archived at https://perma.cc/MZN7-RWPC)

17 Tanker Operator (2024) Red Sea diversions add nearly $1 million per voyage to shipping costs while doubling transit time, Tanker Operator, www.tankeroperator.com/news/red-sea-diversions-add-nearly-1-million-per-voyage-to-shipping-costs-while-doubling-transit-time/14104.aspx (archived at https://perma.cc/56KU-NZ94)

18 Helwa, R and Al-Riffai, P (2025) A lifeline under threat: Why the Suez Canal's security matters for the world, Atlantic Council, www.atlanticcouncil.org/in-depth-research-reports/issue-brief/a-lifeline-under-threat-why-the-suez-canals-security-matters-for-the-world/#:~:text=The%20193%2Dkilometer%20canal%20connecting,trillion%20in%20goods%20transiting%20annually (archived at https://perma.cc/34LT-JVMB)

19 Pereira, I (2025) Trump tariffs won't entice companies to expand US manufacturing, economic experts warn, ABC News, https://abcnews.go.com/Politics/trump-tariffs-entice-companies-expand-us-manufacturing-economic/story?id=120635951 (archived at https://perma.cc/HL9T-SYRR)

20 Espiner, T (2025) Jack Daniel's hits out at Canada pulling US alcohol, BBC, www.bbc.co.uk/news/articles/cn0422zzpw8o (archived at https://perma.cc/2527-N9F7)

21 UN (1926) Slavery Convention, United Nations Human Rights, United Nations, www.ohchr.org/en/instruments-mechanisms/instruments/slavery-convention (archived at https://perma.cc/9YF4-AVX4)

22 Lindsey, H (2019) Use of child labour on cocoa plantations in Ghana, BBC, www.bbc.co.uk/programmes/p07tcv6k (archived at https://perma.cc/5DR8-AS93)

23 Ashcroft, S (2023) Top 10 supply chain sustainability platforms, *Supply Chain Digital*, https://supplychaindigital.com/sustainability/top-10-supply-chain-sustainability-platforms (archived at https://perma.cc/8PM9-282N)

24 United Nations (2013) UN report: one-third of world's food wasted annually, at great economic, environmental cost, United Nations, https://news.un.org/en/story/2013/09/448652 (archived at https://perma.cc/SW37-KWT3)

25 Masterson, V (2022) Here's why food stores are ditching 'best before' labels, World Economic Forum, www.weforum.org/stories/2022/08/waitrose-scrap-best-before-dates-cut-food-waste/#:~:text=Tomatoes%2C%20apples%2C%20potatoes%20and%20pears,of%20apples%2C%20potatoes%20and%20broccoli (archived at https://perma.cc/FPB5-9WGF)

26 Houlton, C (2024) A third of Brits throw away food solely on 'best before' dates, *Grocery Gazette*, www.grocerygazette.co.uk/2024/04/18/food-waste-best-before-date (archived at https://perma.cc/GEV9-S9VE)

27 Coffey, H (2024) Scrap sell-by dates – and say hello to the revolutionary smart food packaging of the future, *Independent*, www.independent.co.uk/life-style/smart-packaging-food-waste-expiry-dates-best-before-b2554649.html (archived at https://perma.cc/R6CK-JL7F)

28 Kroet, C (2024) New EU rules on common USB-C chargers for electronics to come into force, *Euro News*, www.euronews.com/next/2024/12/27/new-eu-rules-on-common-usb-c-chargers-for-electronics-to-come-into-force (archived at https://perma.cc/HES3-GPVD)

29 Product Compliance Support (2024) EU enforces USB-C universal charger rule to reduce waste and save consumers money, Product Compliance Support Ltd, www.productcompliancesupport.co.uk/latest-developments/eu-enforces-usb-c-universal-charger-rule-to-reduce-waste-and-save-consumers-money (archived at https://perma.cc/UGR2-BKF3)

30 Shoaib, M (2023) Meet the brands setting a new blueprint for supply chain traceability, *Vogue Business*, www.voguebusiness.com/story/sustainability/meet-the-brands-setting-a-new-blueprint-for-supply-chain-traceability (archived at https://perma.cc/U8T2-S2LM)

31 Stratten, J (2019) How Net-a-Porter is rethinking retail, Insider Trends, www.insider-trends.com/how-net-a-porter-is-rethinking-retail (archived at https://perma.cc/KKC5-RXKM)

32 Stryker, C and Scapicchio, M (2024) What is generative AI? IBM, www.ibm.com/think/topics/generative-ai (archived at https://perma.cc/5RJL-BT76)

33 Apotheker, J, Duranton, S, Lukic, V, De Bellefonds, N, Iyer, S, Bouffault, O and De Laubier, R (2025) From potential to profit: Closing the AI impact gap, BCG, www.bcg.com/publications/2025/closing-the-ai-impact-gap (archived at https://perma.cc/J7JF-JR7U)

34 Lin, P (2024) Amara's Law and its place in the future of tech, *IEEE Computer Society*, www.computer.org/publications/tech-news/trends/amaras-law-and-tech-future (archived at https://perma.cc/RX5F-E63U)

35 Mowat, M, Butner, K and Sanghera, A (2024) The intuitive supply chain, IBM, www.ibm.com/thought-leadership/institute-business-value/en-us/report/supply-chain-ai (archived at https://perma.cc/J92D-MUP3)

36 Malhotra, G and Abielmona, A (2025) How to transform global supply chain operations with agentic AI, EY, www.ey.com/en_us/insights/supply-chain/revolutionizing-global-supply-chains-with-agentic-ai (archived at https://perma.cc/B5GY-SZLT)

37 Nawaz, S (2024) Top 20 agentic AI use cases with real-life examples, AMPCOME, www.ampcome.com/post/top-20-agentic-ai-use-cases-with-real-life-examples#:~:text=It%20enhances%20safety%20assurance%20by,routes%20and%20reducing%20travel%20times (archived at https://perma.cc/Z2PS-D2VH)

38 Alicke, K, Rachor, J and Seyfert, A (2016) Supply Chain 4.0 – the next-generation digital supply chain, McKinsey & Company, www.mckinsey.com/capabilities/operations/our-insights/supply-chain-40--the-next-generation-digital-supply-chain (archived at https://perma.cc/LC8K-YE2Q)

39 Carvallo, C (2024) My supply chain planners don't want to change, Gartner, www.gartner.com/en/supply-chain/insights/power-of-the-profession-blog/my-supply-chain-planners-dont-want-to-change (archived at https://perma.cc/7WAJ-FBN9)

40 Stilinksi, D, Doris, L and Frank, L (2024) Leveraging generative adversarial networks (GANs) for supply chain optimization: A comprehensive overview, EasyChair Preprint, https://easychair.org/publications/preprint/78ph/open (archived at https://perma.cc/8KJE-RDC9)

41 Metz, C (2024) Quantum computing inches closer to reality after another Google breakthrough, *The New York Times*, www.nytimes.com/2024/12/09/technology/google-quantum-computing.html (archived at https://perma.cc/2MR4-UB7E)
42 Schneider, J and Smalley, I (2024) What is quantum computing? IBM, www.ibm.com/think/topics/quantum-computing (archived at https://perma.cc/YB8D-8DPT)
43 Milmo, D (2025) UK cybersecurity agency warns over risk of quantum hackers, *The Guardian*, www.theguardian.com/technology/2025/mar/20/uk-cybersecurity-agency-quantum-hackers (archived at https://perma.cc/UJ4K-9B8A)
44 Execs In The Know (2023) The rise of environmentally conscious consumers: CX reimagined for sustainability, *Execs In The Know*, https://execsintheknow.com/magazines/july-2023-issue/the-rise-of-environmentally-conscious-consumers-cx-reimagined-for-sustainability/#:~:text=Companies%20can%20cultivate%20trust%20and,brand's%20sustainability%20efforts%20before%20purchasing (archived at https://perma.cc/V2F9-WA29)
45 Junaid, M, Zhang, Q, Cao, M and Luqman, A (2023) Nexus between technology enabled supply chain dynamic capabilities, integration, resilience, and sustainable performance: An empirical examination of healthcare organizations, *Technological Forecasting and Social Change*, 196, November 2023, www.sciencedirect.com/science/article/abs/pii/S0040162523005139?via%3Dihub (archived at https://perma.cc/76T7-JM48)

PART FOUR
Conclusions

Key takeaways 14

Leveraging technology for competitive advantage in supply chain management

This chapter provides a summary of the key takeaways from the book, including the importance of technology in driving competitive advantage.

> **CHAPTER AIMS**
>
> - Understand the importance of technology in driving competitive advantage.
> - Recap the key takeaways from the book.

'You cannot stay in one place: you either go forward or you go backward.'

THOMAS WATSON SR[1]

As we conclude this journey through the supply chain and the role that technology is playing in modernizing companies to meet and exceed customer expectations, as well as amplifying the importance of the topic in the boardroom, there are many themes that we need to bring together.

Through this journey, my intention has been to shine a light on technology and how it is an enabler for leaders to keep control of an ever-expanding, complex, 'always on' supply chain and assist them in driving competitive advantage. Customer expectations and global market trends mean that supply chains cannot stay static or they will become uncompetitive, which makes Watson's quote above appropriately fitting to describe modern supply chains.

The advancement of technology is not slowing down and while you can feel ignorant rather quickly if you don't keep on top of it, the real focus should be on how to balance getting your current supply chain performing as well as it can, at a cost which doesn't compromise incremental investments to maintain its ability to provide an advantage and aid the company in being successful. The days of investing in systems that take multiple years to implement, let alone gain benefits from, will not be useful. Building entire applications on underlying systems that are not fit for purpose, while easier to get approved, are also not useful; they are doing nothing but complicating the spaghetti IT architecture landscape that many companies have. As I outlined in Chapter 3 and developed in Chapter 13, this need for a solid baseline is critical, and determining how to get to that place at speed is important since it will get you positioned for the capabilities that will drive advantage that are enabled from this base. Supply chains need a strategy: yes, it must align with the business but just as importantly, it must be clear and concise on what capabilities it wins with today and what it requires in the future, from people, process, technology and data perspectives.

The key to answering the question of 'how will the technology drive competitive advantage?' is to break down the elements that the supply chain can influence through its performance and determine why technology, working side by side with employees, enhances this. However, too many times, companies look at the specific problem they are trying to solve in isolation and don't consider the overarching architecture. The technology solution might solve that specific problem but then cause problems elsewhere or shine a light on other areas that are now inefficient. This is where a solution that can neatly fit into the existing architecture and be implemented quickly should be chosen, rather than one which leads to re-plumbing much more ground of the supply chain which takes us back to the long, expensive implementation.

Be clear on what you want from the technology before engaging with vendor solutions – don't engage too early either. The number of times I have heard that a software has been chosen without the company documenting its business requirements or that has been influenced by a great salesperson who has promised their solution can do everything in a specific domain, sadly no longer surprises me.

Once the technology required to assist the company in its mission has been decided, the business case must be outlined. This should not be a one-off exercise; it should evolve from an initial idea into what the technology is needed for and what positive impact it will have on our customers, colleagues and overall business success. Companies have very different approaches here; however the key pillars are having owners responsible for delivering the project, the benefits it will provide (both tangible and intangible), timeline, risks and the costs. It can be quite transactional at this stage, so supply chain leaders need to balance the development of the case with winning over hearts and minds of influencers and decision makers, with a strong plan of how to move forward being important in the context of the company overall and its mission.

The first place to start is in relation to efficiency and cost.[2] Companies should be thinking about their processes and how they can make them intelligent by automating those that are manual and time-consuming for users. This should be reviewed at the process level because automating a workflow in silo may provide an improvement, but it is at the process level where greater gains can be achieved. For example, if looking at a workflow, it might make sense to automate a task such as a user having to click 'accept' on a screen, then go into another screen and then click 'accept' there as well. However, if you look at the overall process, it might make sense to put in further steps of automation rather than just what one user does. This can be achieved by documenting users' key workflows and then looking at which parts are cumbersome and slow. The overall goal is to reduce the manual activities and have a rigorous workflow in place, which frees the user to focus on value-adding activities where human decisions need to be made. By following this approach, the company and the users will create greater levels of visibility, highlighting where technology can do more heavy lifting for users and then implement the changes. This does not require significant investment in technology. Tools such as data mining and robotic process automation[3] can, in many cases, help to solve or be part of building fact-based reasoning to why a larger change driven by technology should be pursued (e.g. implementing a new planning system that will cover many workflows). The main outcomes of this work, beyond assisting users, are to increase efficiency, productivity and reduce costs.

Technology that aids the supply chain

We now turn to the fundamental technology that underpins the supply chain; it would be amiss not to reflect on these again.

Visibility and transparency

As you have read time and again in this book, visibility and transparency go hand in hand and are a significant unlock for not only the benefits outlined previously but also in trying to move the dial when it comes to gaining advantages. There are many solutions available to companies with excellent use cases available in the public domain and/or through system integrators.[4] The key technology components for this are cloud computing, IoT, blockchain, digital twins, data analytics and advancements in AI. When looking to improve visibility of the supply chain, the initial challenge is to determine what technology capabilities you have currently and how these, stitched together with newer technology, provide the answer or set the company on the journey to seeing its supply chain through a technology lens. In some extreme cases, it may highlight a company's need to reset its technology stack to then move forward. Here, the question of how much visibility it needs to have can be addressed. This is a dilemma for supply chain professionals because many will say that higher visibility is better. However, this can lead to a lack of focus. Time may be better spent on getting their part of the supply chain performing as best it can.

Also, with there being so much ambiguity day to day in the supply chain, supply chain professionals must ask whether having that visibility adds value to the company? Being able to act on the findings gained from having a higher visibility should lead to competitive advantage; however, the level of visibility is subjective to company needs. Being able to know at what stage the inventory is at, whether raw material, work-in-progress or finished goods, is crucial for companies and is an obvious given but many don't have this information to hand, minute to minute. That might be good enough, but it is dependent on the type of operation and speed the supply chain needs to respond to.

Then, considering the steps from the inventory leaving the warehouse as finished goods to reaching the customer, how often does the user or customer need to be made aware of where the shipment is at any time? We can have notifications to inform us at each stage of the delivery journey. Questions like these need to be asked and then the technology developed to support the desired outcomes. When a retailer asks these questions, it might be more critical for them to have that visibility and to share it with the end customer to build and maintain trust, versus a consumer products company or a supplier to a consumer products company. When considering transparency, similar questions must be asked. For example, shipments will be more focused on when ownership changes and then making sure transactions between parties are executed at appropriate times.

When looking through a consumer lens, they may want to know what their clothes are made of and where they are made. However, do they see the value of that investment from the company providing this information and are they willing to pay for that knowledge? Is the company happy to make the investment in technology to provide this? It is these trade-offs, as well as the brand strategy, that should guide the decision. When looking from customers back upstream, how much information and at what time intervals do all the parties involved need to have it to make informed decisions should be evaluated and not just assumed that second by second is worthwhile at each stage.

Data and analytics

For technology to make a difference, how a company looks after its data is a key enabler. Data can be kryptonite for companies, with too many years of treating it poorly making it increasingly challenging to get to insights to drive business decisions.[5] To stop moving forward with technology because of poor data is not the answer, but finding the pockets of data that are good enough should be prioritized. Then, this can be used to develop capabilities on top of the existing technology and improve the data in conjunction. If you hear colleagues saying we will tidy up the data first, please challenge them. One person's definition of good enough is poor to another, so getting clarity on the quality of data and what 'better' looks like needs to be agreed quickly. The key point is that without having a data-centric mindset in supply chain, pursuing technology capabilities will not derive the expected benefits.

By using advanced analytics, such as predictive and prescriptive analytics, companies can transform raw data into actionable insights to improve areas like planning, supplier performance monitoring and transportation optimization, ultimately reducing costs, enhancing customer satisfaction and providing a sustainable competitive advantage.

Role of automation and robotics

Technology has been seen to work well in tandem with people in warehouses. Automating the pallet-in / pallet-out operation has been most effective and has led to the reduction of warehouse roles in those types of operations. With the rise of e-commerce requiring speed of throughput, including single picks, this still necessitates many warehouse roles to work with technology to complete tasks.[6] For example, Amazon ramps up demand for staff during busy periods, despite the evolution of technology solutions.[7] With more automation in warehouses, the ability for companies to understand key metrics in relation to performance has led to benefits using analytics to predict effort versus demand and to innovate with new technology to better aid the user. For example, cobots are an enhancement that helps reduce the effort and time users need to spend looking for products aligned to orders. Many warehouse technology investments are significant and the payback period can be quite long, but these investments are expected to last over 10 years[8] (depending on the warehouse design and expectations of the warehouse being able to handle capacity predicted in the medium term). Having automation in their warehouse operations will allow companies to offer later cut-offs and execute tasks outlined over a shorter time horizon.

Transport technology can be used to orchestrate transportation, providing companies the ability to define routes and get the best price from carriers competitively by using systems that carriers can bid on in near real time. Depending on the company and what it does, optimizing routing and providing customers flexibility on where they can receive orders, whether that's to stores, lockers, homes or offices, aligned to delivery dates that meet or exceed expectations, can also be a differentiator. Then the customer experience, such as providing them with tracking information in near real time and personalized notifications, is a nuance that is likely to lead to improved customer satisfaction, loyalty and retention, which is fast becoming 'table stakes' in capturing growth in markets such as e-commerce. This area has embraced the use of technology and continues to be at the forefront with investments in using AI to find improved ways to service customers both from physical and information perspectives.[9]

Digital platforms

Platforms enabled by the cloud have helped advance the use of technology and data accessibility both internally and with external partners, enabling better collaboration and integration across global supply chains and therefore improving transparency, efficiency and flexibility.

Successful adoption of digital platforms requires clear objectives, stakeholder buy-in and collaboration, and ongoing optimization to ensure alignment with supply chain strategies, especially in volatile environments.

Technology that runs current supply chains

Enterprise resource planning (ERP) systems are a great place to start. They provide the foundations, setting the tone for data and its use and the record of transactions for the supply chain. However, they alone don't offer a competitive advantage. Trying to fit a system as complex as an ERP onto bespoke processes,[10] which an individual company may have is a struggle (as we have seen in the case of Lidl and there are many others)[11] and trying to stay in a repeat loop of upgrading is exhausting. Even the sellers of these solutions encourage a standardized approach to processes across the functions of a company, which only emphasizes that it is 'vanilla'[12] (i.e. sameness). Vendors since the late 2010s have been selling supply chain modules separately,[13] which suggests they are beginning to understand that all-encompassing systems are not the answer for next-generation supply chains, since they point to basic functionality and do not provide the organization with a competitive advantage. The implementations can go over budget and in some instances, never finish. Investors and other interested stakeholders need to see tangible benefits at regular intervals to support significant investments.[14]

A standardized approach makes sense for the execution of non-value-adding activities,[15] but when our attention turns to value-adding activities that bring companies success, ERP systems do not provide this for the supply chain. However, companies that are global and large in size need a good, strong ERP because it provides the transactional elements that need to be managed and for compliance purposes; it is part of the company DNA and it is an enabler for other capabilities, so it can feel like a 'Catch 22' situation.

Planning is a major decider of how well a company will perform in its supply chain and is very difficult to get right consistently.[16] Having point of sale (PoS) information is helpful to actors across the supply chain but needs to be treated carefully. If you are the supplier and are reacting to daily changes in sales, the messages you are acting on and then signalling to your supply base and beyond could lead to the 'bullwhip effect.'[17] (where the reaction is amplified, the further back you are in the end-to-end supply chain). This could lead to significant over- / under-stocking and is one of the fundamental issues that supply chain faces. Having strong planning capabilities is a driver of modern technology enhancements, where the landscape of vendors who offer solutions has dramatically increased since the late 2010s.[18]

Where companies need to tread carefully is how do their supply chain professionals want to plan and how can technology help them achieve this? Many demand models are built fundamentally on mathematics that were developed in the mid-20th century and still stand the test of time today, with the real advancements coming from the ability to introduce different variable sets to this to inform forecasts. Also, although AI is driving improvements in forecasting, to say AI is the answer is inaccurate. A planner still needs to assess the situation and all the extra data that the different variable sets provide, and work with commercial colleagues to guide what they think provides the most accurate view of the products and categories. Forecasting will always remain a subjective area and as we discussed in Chapter 5, don't forget forecasts are inherently wrong, so to suggest waving an AI wand will make forecasts perfect is a fantasy.[19]

Also, in conjunction with this, many planning technology solutions still encourage users to engage with spreadsheets and then upload these to the system. These should be built into the application to help reduce the time users waste on moving between screens. One further point to add here is transparency on why the maths embedded in the technology is stating what it is in relation to a forecast. If a user sees a number they don't agree with, it can quickly escalate to a trust issue. The user then spends time out of the system on spreadsheets to come up with the answer they believe is correct, wasting time that the system was meant to reduce. Therefore, a lack of transparency can encourage users to do much more out of the system rather than working in concert with it, where both parties learn from each other to propose the best forecast based on the information they know. Having additional capabilities, such as forecast value-added (FVA) analysis, available with other analytics capabilities, will help the user quickly grasp what is happening and therefore be informed in the planning process and ultimately drive benefits for the company. Scenario planning has been a great addition to these systems and this gets to the crux of value-adding activities. Being able to quickly evaluate situations and make informed decisions reduces the lag time and helps elevate the supply chain to being proactive when facing uncertainty.[20]

Advanced planning systems, beyond giving the planner much more data and insights to work from, also help the supply chain optimize inventory positions, which in turn will lead to a reduction in stockouts and reduce waste associated with too much inventory, offering a competitive advantage. Planning is an undervalued skill that technology can most certainly help users with.

However, what the current technology solution can or cannot provide to the user is usually found out post-implementation. Providing the user with technology that helps them gain the most functionality is still not available. There are elements in the 'advanced planning' systems that are useful, but they usually come with a clunky user interface, which is not intuitive. It would be more advantageous for them to operate with natural language programming, where the user can ask the system questions to aid them rather than having to search through countless screens to review and make decisions from. This could then, in time, lead to a write-back capability as discussed in Chapter 13, where decisions made at the user interface level go back into appropriate systems instead of the user having to do all that work. This would advance the impact and benefits these systems would have with users and I expect in due course that the next versions of 'advanced planning' systems will have capabilities like these as standard.

Technology is aiding companies with getting a much stronger handle on the supply base they manage and helping users be more intentional with the company's expenditure. Many technological advancements are taking the core principles of procurement and its workflows and infusing them with new capabilities to drive better outcomes for users. Earlier in my career, the collation of suppliers and their spend would be an exercise in itself to compile data, such as the terms of contract and knowing the location of where raw materials are made to identifying new potential suppliers. Now, this has rapidly evolved to a point where companies that have implemented technology in sourcing can get to relevant information on suppliers and their performance with only a few clicks. They can view category evolvement and spend to prioritize category strategies quickly.

However, where technology is still struggling in this area is moving beyond structured traditional ways of procuring, like using advancements in AI to aid users in their roles. Sourcing technology is most efficient when it can recommend alternative approaches and even execute purchases based on guardrails.

Other helpful tools to identify procurement inefficiencies and provide solutions for them using technology include data mining.[21] This finds ways to streamline and improve strategic activities in relation to source contract processes and improve operational activities, primarily in relation to purchase-to-pay (lifecycle of purchasing goods from identification to payment) processes. When companies get these working strongly with technology, the benefits of speed, onboarding and leveraging spend drive excellence and productivity into the process.

Resilience is another component to consider, specifically looking upstream to get visibility and control of the supply base and risks. Where the responsibility for resilience sits in a company may differ based on what it is in relation to but within the context of a manufacturer, a significant risk to its ongoing success is a lack of resilience through the execution of the procurement function.

Resilience has come and gone as a trend that supply chain leaders focus on, but an improved technological ability to support it has meant it is now a priority. Resilience technology solutions can aid companies in identifying and mitigating potential risks in the supply chain (e.g. natural disasters, supplier insolvency, child labour) and provide analytics and simulation tools to help users model scenarios[22] leading to the development and execution of contingency plans. With technology providing the ability to collate and digest vast amounts of data from a variety of sources, resilience information has become more available and has been used more to inform decisions. It assists companies in making proactive decisions and helps influence their supply base to make changes to support the overall resilience of the end-to-end supply chain, such as adjusting manufacturing of key raw materials within a supplier and having dual sourcing strategies to help reduce risks.

Measuring a supplier's resilience score from various metrics, such as where they operate in the world, the political stability of a country or region and the journey products take from manufacturer to customer, can help companies make informed decisions on how they want to operate. If there is a conflict in a specific region, knowing this early can reduce the risk of failure to customers and provide opportunities to move products via alternative routes or turn to alternative suppliers.

It can be difficult to put a financial value on the benefits when devising a case for implementation of resilience technology capabilities since they become part of the day-to-day activities of users and are used in conjunction with other systems to manage the supply chain. They have benefited companies across various sectors in managing risk. Financial benefits can be linked to reductions in freight expediating, reductions in premiums paid for direct material purchases, reductions in inventory carrying costs and time to resolve supply risk issues.[23]

We have not spent considerable time discussing manufacturing within the book primarily because advancements in this space have been embraced more than in other areas of the supply chain, with technology already making a considerable difference to operations.[24] The use of technology to run production lines has been common practice since the industrial revolution.[25] Innovations include reviewing performance from an analytics perspective, with trial and error to optimize throughput where possible. To evaluate the efficiency of machinery, predictive maintenance has been at the forefront. The technology, typically a combination of IoT and analytics working in tandem, helps gauge when maintenance should happen, helping the overall supply chain plan proactively for this and to reduce downtime and the risk of failing customers. Digital twins (discussed in Chapter 4) have also been deployed to help better understand equipment, to monitor performance and to run scenarios, where minor changes to the operation are made and the impact assessed. Embracing technology to enhance manufacturing has reduced the number of workers required and has led to significant improvements in health and safety. Also, this technology has improved training users by being able to codify operations and use versions of digital twins, such as augmented reality and virtual reality glasses to accelerate upskilling, which has led to reducing the time from hiring to working in production.[26]

Conclusion

The main takeaway is that technology in the supply chain does help companies with competitive advantage when deployed in conjunction with people, process and data. The focus of the technology solution should be on the customer predominantly to begin with because with them in mind, it will prevent you from the distractions of operating with a 'plasters' mentality. The company should discern what service it believes will provide it with the edge over competitors and establish it as a leader in the marketplace, which leads to improved market share, revenue growth and brand recognition. Then that solution (considering the mix of elements) should help design a supply chain that has agility and flexibility to meet customers' expectations and evolve over time.

The juggling of trade-offs aligned to the pentagon while trying to respond to changing customer expectations is a very large challenge that will always be at the forefront of companies' minds. However, bringing it back to first principles, providing a good service and quality of product is non-negotiable and that should guide the decisions you make in relation to supply chain technology investments.

With changing markets and the pace of change with technology, companies can be fearful of which actions to take regarding the supply chain. Supply chain professionals must remember, though, that there are reasons why the key capabilities that underpin supply chains are still in place 30 years later. Yes, there are many complexities, but the basic, underlying processes of obtaining a product from raw materials, packaging and sending it to a customer, are essential to execute. Technology and the innovations appearing before us now, provide the ability to see into the supply chain and make more informed decisions with data and insights in near real time, a position which was never possible before.

Endnotes

1. Maney, K (2007) *The Maverick and His Machine: Thomas Watson, Sr. and the Making of IBM*, Wiley
2. IBM Institute for Business Value (2025) IBM, IBV Research, IBV Performance Data and Benchmarking Program, 2024, IBM, www.ibm.com/thought-leadership/institute-business-value/en-us/benchmarking (archived at https://perma.cc/K999-LHMF)
3. IBM (2025) Process mining and RPA: The automation power couple, IBM, www.ibm.com/resources/automate/maximize-process-mining-with-rpa (archived at https://perma.cc/6ZKS-KECF)
4. Bailey, E (2025) Why better supply chain visibility is critical in 2025, *Supply Chain Dive*, www.supplychaindive.com/spons/why-better-supply-chain-visibility-is-critical-in-2025/735993 (archived at https://perma.cc/4MV5-9KQ7)
5. The Sellbery Team (2025) Why most companies are still struggling to use their data effectively, Sellbery, https://sellbery.com/blog/why-most-companies-are-still-struggling-to-use-their-data-effectively (archived at https://perma.cc/QH8F-6FW9)
6. Ashcroft, S (2024) How automation tech is easing warehouse labour problems, *Supply Chain Digital*, https://supplychaindigital.com/articles/how-automation-tech-is-easing-warehouse-labour-problems (archived at https://perma.cc/48RF-WGQ3)
7. Delesline III, N (2024) Amazon wants to hire 250K for the holidays again, *Retail Dive*, www.retaildive.com/news/amazon-hire-250k-holiday-season/728820 (archived at https://perma.cc/2ZFD-YVK5)
8. Davies, A, Oca, A, Prieto, I and Sarilar, U (2023) Getting warehouse automation right, McKinsey & Company, www.mckinsey.com/capabilities/operations/our-insights/getting-warehouse-automation-right (archived at https://perma.cc/2ZFD-YVK5)
9. Cook, R (2024) Enhancing supply chain resilience through logistics technology, Supply Chain Brain, www.supplychainbrain.com/blogs/1-think-tank/post/38893-enhancing-supply-chain-resilience-through-logistics-technology (archived at https://perma.cc/3LQY-Y35E)
10. Braun, F and Östhed, H (2019) Rethinking your ERP strategy to gain digital competitive advantage, Bain, www.bain.com/insights/rethinking-your-erp-strategy-to-gain-digital-competitive-advantage (archived at https://perma.cc/87M8-6UXV)
11. Lawton, G (2024) 10 notable ERP implementation failures and why they failed, TechTarget, www.techtarget.com/searcherp/feature/7-reasons-for-ERP-implementation-failure (archived at https://perma.cc/LF7H-977Y)
12. Bird, A (2022) Handling complexity: should we be scared of standardisation? ERP Today, https://erp.today/handling-complexity-should-we-be-scared-of-standardisation (archived at https://perma.cc/RXH6-ZMYZ)

13 Jackley, M (2023) 10 Top ERP modules and their features, Oracle, www.oracle.com/uk/erp/erp-modules/#:~:text=ERP%20functionality%20is%20organized%20by,collected%20into%20a%20central%20repository (archived at https://perma.cc/KC92-QKST)

14 Dolfing, H (2020) Case study 6: How Revlon got sued by its own shareholders because of a failed SAP implementation, henricodolfing, www.henricodolfing.com/2019/08/case-study-revlon-sued-because-sap-implementation.html (archived at https://perma.cc/T8GK-CM7A)

15 McCue, I (2025) 15 benefits of ERP for businesses in 2025, Netsuite, www.netsuite.com/portal/resource/articles/erp/erp-benefits.shtml#:~:text=Standardized/Centralized%20Data,time%20and%20money%20as%20possible (archived at https://perma.cc/YKT7-REML)

16 Skill Dynamics (2025) Demand forecasting for smarter supply chain planning, https://skilldynamics.com/blog/demand-forecasting-supply-chain-planning (archived at https://perma.cc/7KWN-8Z64)

17 Sheffi, Y (2022) Prepare for the bullwhip's sting, *MIT Sloan Management Review*, https://sloanreview.mit.edu/article/prepare-for-the-bullwhips-sting/ (archived at https://perma.cc/WU7Y-EACF)

18 Orup Lund, P, Graham, J, Thomson, C, Brett, S and Dawkins, E (2025) Gartner magic quadrant for supply chain planning solutions, Gartner, www.gartner.com/en/documents/6356179 (archived at https://perma.cc/6YV8-DGQ4)

19 McDonald, P (2021) Demand forecasting is always wrong: Three ways to thrive with forecast ambiguity, Forbes, www.forbes.com/councils/forbestechcouncil/2021/04/27/demand-forecasting-is-always-wrong-three-ways-to-thrive-with-forecast-ambiguity (archived at https://perma.cc/PLN9-SZXR)

20 Joglekar, N and Phadnis, S (2020) Accelerating supply chain scenario planning, *MIT Sloan Management Review*, https://sloanreview.mit.edu/article/accelerating-supply-chain-scenario-planning (archived at https://perma.cc/3WTH-Y6GH)

21 Couchbase Product Marketing (2024) What is data mining? Techniques, tools, and applications, Couchbase, www.couchbase.com/blog/data-mining-techniques/#:~:text=Data%20mining%20tools%20are%20software, interactively%20handle%20large%20data%20sets (archived at https://perma.cc/D6ME-WL5Y)

22 Te Lindert, M (2025) Scenario planning as a tool to turn risks into opportunities, *Supply Chain Movement*, www.supplychainmovement.com/scenario-planning-as-a-tool-to-turn-risks-into-opportunities (archived at https://perma.cc/L2C8-5VM3)

23 Li, P, Chen, Y and Guo, X (2025) Digital transformation and supply chain resilience, *International Review of Economics & Finance*, 99, www.sciencedirect.com/science/article/pii/S1059056025001960 (archived at https://perma.cc/YX2G-HL72)

24 Huang, K, Wang, K, Lee, P K C and Yeung, A C L (2023) The impact of industry 4.0 on supply chain capability and supply chain resilience: A dynamic resource-based view, International *Journal of Production Economics*, 262, www.sciencedirect.com/science/article/pii/S0925527323001457#:~:text=Industry%204.0%20indicates%20the%20revolution,et%20al.%2C%202017b) (archived at https://perma.cc/5Q8L-UU84)

25 Schwab, K (2016) The fourth industrial revolution: What it means, how to respond, World Economic Forum, www.weforum.org/stories/2016/01/the-fourth-industrial-revolution-what-it-means-and-how-to-respond/#:~:text=The%20First%20Industrial%20Revolution%20used,%2C%20digital%2C%20and%20biological%20spheres (archived at https://perma.cc/4VR9-FZYE)

26 Wichienwidhtaya, J (2021) Utilizing VR for manufacturing learning and training, Western Washington University, https://cedar.wwu.edu/wwu_honors/495 (archived at https://perma.cc/8BH5-7AKZ)

Final thoughts and future considerations 15

This chapter will cover the final thoughts and future considerations for supply chain leaders, including the importance of continuous innovation and improvement.

CHAPTER AIMS

- Embrace the importance of continuous innovation and improvement in supply chain management.
- Learn why ongoing learning and development are essential for staying competitive.

'The most effective way to do it is to do it.'

WIDELY ATTRIBUTED TO AMELIA EARHART

REAL-WORLD EXAMPLE Bell Labs

Bell Labs is a research and development company renowned for its innovations. It created many products and concepts, such as the transistor and information theory (a foundational work for modern technological advancements), which is now a subsidiary of Nokia. Many of its approaches to innovation that made it successful should be considered within the supply chain community to help foster innovation and improvement, as well as ongoing learning and development.

It disappeared from the front pages due to the 'telecoms crash' in 2001, when global stock markets crashed due to unrealistic expectations of growth and profitability.[1] At the time, companies like Enron and WorldCom were found to have intentionally misled markets and shareholders, putting them both into bankruptcy.[2,3] Other companies lost much of their value and then had to reposition themselves, among them Bell Labs, which was spun off from AT&T. Bell Labs became part of Lucent Technologies, which merged with Alcatel and became part of Nokia in 2016.[4]

It continues to be at the forefront of technology, exploring AI, quantum computing and 6G wireless technology.[5]

Its success, particularly from the 1920s to the 1980s, can be owed to the culture it fostered, which encouraged and rewarded experimentation, collaboration and continuous learning. Employees with diverse backgrounds worked through ideas, sharing their knowledge and expertise. Employees were given freedom and time to explore new ways to solve problems, which were applauded even when they led to failure (which was seen as a crucial part of learning). They were also early users of the mentoring approach that formalized the sharing of knowledge from experts to junior colleagues. They developed and sustained a culture that embraced the unknown and were able to balance theory and practice, which helped them move from conception to prototypes quickly because they were clear on the objective and bought into it. Their success was for the team, not the individual.[6]

For supply chain professionals, the key takeaways from this book should be to get out of your silos and start to understand what your role contributes to the overall ecosystem and how your actions affect other parts of the supply chain. Experiment with technology and challenge the activities that you do daily: is there a better way? How can you get to the information you need rather than spend countless hours in and out of different applications? Technology should be your friend; look to implement technology that eliminates the need for additional manual steps or having to do a workaround for an activity and focus on turning data into trustworthy data – you need to be a catalyst and an optimist in this endeavour.

Although we have reached the end of this book, there is so much more to say on supply chain and how far it has come as a profession. The combination of people, process, technology and data has driven so much change and has an exciting future ahead.[7] Supply chains need to be appreciated for what they are: the backbone of a company.[8] When it is stressed, it can have major ramifications on a company's bottom line and the perception of the brand in the marketplace – one just needs to look at the challenges during Covid[9] that I have outlined in previous chapters.

Throughout the book, reference has been made to people, process, technology and data and the need for these to be orchestrated as one to achieve the goals of the company. This is never more the case than when examining the supply chain. We have gone through the main subfunctions of a traditional supply chain and while there are new technologies arriving frequently, what we must remember is that fundamentally, all elements of the supply chain must work in sync when trying to fulfil a customer order to achieve the overall mission of delighting the customer.

We don't operate in our own bubble

Supply chains love stability, predictability and control. However, they cannot operate in their own bubble and are exposed to what is going on at local, regional, country, continental and global-level factors that are beyond their control in many instances. We have delved into many factors (e.g. inflation, wars, tariffs) and while some of these issues may disappear as quickly as they arrived, there will always be factors affecting the supply chain. Whatever variables are affecting them, we need our supply chains to be responsive and find ways to innovate to keep being competitive in the market. Where it can be difficult is in prioritizing supply chain technology investments. For example, they can be deprioritized in the short term due to companies having to absorb lower margins to remain competitive as their costs increase or where changes in leadership roles might elevate another C-suite member's agenda. Whatever the reason, we need to be compelling in how we explain what the technology provides and how it connects to improved outcomes for our customers and the company's performance.

The emergence and rise of online shopping has elevated the role of the supply chain further, with refreshing physical networks to be able to move products through quicker to the customer who wants everything immediately and the technology that supports it.[10] Order portals linked to order management systems that subsequently connect to inventory records, warehouse management systems and carrier systems have been a must for any company wanting to fulfil this customer channel.

The immense work of people like Al Gore, the founder and chairman of The Climate Reality Project, showed us what our actions are doing to the global climate and the ramifications back in 2006,[11] which focused our attention on the environment. Yet, in 2025, certain government agendas[12] suggested dialling back on sustainability plans. Despite this, we have technology in place with analytics capability to be able to move from simply surfacing data to being able to action it. The biggest challenge to this is moving from estimates of emissions data to actual emissions data, which we need to find a solution for. This is likely to be achieved through using AI within the process as outlined in Chapter 13.

The themes covered in this book have a direct influence on what we do in the supply chain, and we should not lose sight of how these are in dichotomy to stability, predictability and control of supply chains that we so crave. Therefore, we must face them and find ways through people, process, data and technology to help address them as best we can.

Technology is our friend

Technology has had significant implications on supply chains, not just from specific capabilities to run them but also by influencing what type of service consumers experience, such as being able to receive notifications, engaging with chatbots and rescheduling deliveries, for example. The advancements that have led to where we are today from codifying mathematics[13] to the internet to the emergence of artificial intelligence,[14] show that the space is constantly innovating. It is worth considering Amara's Law here – many technologies are applicable in the supply chain and we should be cautious not to overestimate and hype up the technology in the short term and not to underestimate it in the long term.

The supply chain comes to life from a demand perspective, which in a sense drives the rest of the supply chain. Capturing demand shows which elements (i.e. inputs) need to be sourced to transform into a finished good that meets what the customer wants, where and when. Although this sounds simple, orchestrating all these activities with the interaction of people, process, technology and data is complicated.

The role of technology as an enabler has driven much of that innovation.[15] For many years, the goal was to use technology to provide a 'good enough' view of the supply chain, but in most instances, this was like looking in the rearview mirror – the supply chain was dependent on the predictability and stability of each part of the ecosystem, as well as the knowledge and instincts of people to make the right decisions. As we have embraced technology and had to adjust to changing market and consumer expectations, the rearview mirror isn't helpful enough. If we're not careful, an inevitable crash will happen because we're too busy looking everywhere except ahead.

It is fair to say that up to the mid-2000s, the operational element (i.e. physical assets) of the supply chain was predominantly focused on gaining competitive advantage driven by the knowledge and instincts of people and processes. This explains why we accepted the technology provided and were happy to use manual workarounds to make up for its lack of flexibility and for those systems to be predominantly a record of transactions that did not embrace technology and data wholeheartedly.[16]

When the internet emerged, it gave us the ability to connect and share data, which was then accelerated with the introduction and commercialization of IoT devices, whereby surfacing and exchanging data was able to take place everywhere, providing near real time information.[17] Digitalizing user actions has led to a shift in perspective, enabling the growth of supply chain technology to the point now where, in some subfunctions, we have many solutions available to assist users with running their operations. Data is the foundational element; any technology built on poor data is like building a house on quicksand. Within the technology's implementation, there must be activities linked to data and how to refresh it (if required). Also, importantly, how are you going to keep it of a consistently high standard when in business-as-usual mode and make sure errors can be ironed out early? Remember that for technology to be an enabler, it needs quality data but it doesn't need to be perfect. Starting small with good data aligned to the technology you are wanting to implement is key and then in time, those marginal gains will lead to improvements across wider supply chain data.

In the past, we have thought of technology in supply chain as, 'I have a problem: help me put in a system which surfaces data and can answer my questions on how I should solve that.' Implementing technology is as much a trust issue with people as it surfaces fear around change. However, with the way supply chains need to operate today, trust and letting go of repetitive non-value activities is a key to success. We also need to evaluate the processes that drive our activities to an outcome, mapping these and challenging them. Only then can we get into the process workflows and brainstorm ways to mix the four elements (people, process, technology and data) to design what works best for our companies to drive efficiency and improve decision-making.

Innovation overload

A culture of being willing to experiment and test new technologies is required for companies to keep evolving. There are examples of pioneers who are at the forefront, such as Unilever (Chapter 4) and IBM's Infrastructure business (Chapter 13), but the key themes of visibility, transparency, analytics and autonomous decision-making are seen at various levels of maturity in companies across all sectors today. Before implementing technologies that help us achieve maturity, we must be clear on what our supply chain strategy is – this should be what guides us. The strategy does not have to be static. However, avoid having the entire supply chain in constant flux with different pilot projects or looking at problems to solve in isolation. These actions just encourage more and more siloed spaghetti technology landscapes.

We need speed in deploying new technology capabilities in supply chain, so we must challenge the underlying idea of accepting multiple years to implement a solution. If we need to do a transformation of capabilities, how do we split the project up so we can implement capability at specific points along the project, which will deliver benefits regularly, build momentum and excitement to get access to the capability into users' hands as quickly as possible? The answer here is to challenge the 'one size fits all' standardized approach, borne out of the traditional ERP implementation approach and the desire of standardization across all markets. Now, companies should assess what provides other markets with similar functionality, grasp what drives the value and implement this so that benefits can be realized.

Turning to analytics and the need to get to insights quickly has always been a priority for supply chain professionals and technology has helped to make this readily available with limited user input. The ability to see at any one time what is going on, whether in a warehouse or the overall supply chain performance, has been a significant benefit. We have moved from the static nature of predominant analytics usage (the stitching together of KPIs looking backwards), advancing to where we can now gain insights and predictions of what is coming, the 'seeing around the corner' moment and then acting on it and reviewing anomalies.

Looking towards a future that we are now embracing, with generative AI, agentic agents and quantum computing, it is very easy to be a naysayer and hold onto what we know. Inherently, people struggle with change. There are concerns with technological advancements, that cyber-attacks will occur more regularly in the future, putting the security of company technology and data near the top of the C-suite agenda and having to manage this risk proactively. This shows how fragile the supply chain can be even when users are on top of all the other factors at play and is a major area of investment that will need continuous monitoring. We need to also be careful not to think AI advancements alone will provide sustained competitive advantage: the role of creativity and passion from our people will still be at the core of future supply chains.[18]

Generative AI seems to be everywhere, being used for many different activities, but its application to supply chain is in the experimental phase. Advancements such as the ability to analyse and pull out key information in relation to procurement contracts are an excellent use case driving value. Using GenAI to improve the forecasting process is another example. Being able to train a model on multiple datasets and ask it to help identify key information to feed the forecast is plausible and in time this will add value to the process. However, I believe it will bring value only when companies develop their own smaller language models that are trained on their data to imagine their future. This data must be kept refreshed, which is not the case with many large language models deployed today, which use data from a specific time period and are readily available to all – hence the lack of competitive advantage.

Agentic AI is exciting. Having multiple agents to help the user interrogate various datasets to make informed decisions is expected to be mainstream within a few years, with many open sources providing an entry point today. The challenge will be how to codify the activities you want the agent to do: how much freedom you allow them, how trusting the user will be in their outputs and how to overcome any lack of confidence that they do have.

Language models and their vision for supply chain working with GenAI and agentic AI will need to move from large language models to small models (enterprise-focused), then scale up to medium models (enterprise plus collaborators) to gain traction. This is because in 2025, companies were still chasing to condense the size and cost of large versions. The real breakthrough for supply chains will be when we feed them with our own data.

Quantum computing needs to be commercialized at a price point companies can afford and like other technologies, will need to be explainable to gain users' buy-in. The ability to analyse multiple options simultaneously will be a step change for companies, allowing them to optimize their supply chains in near real time. However, this will require a different approach to how supply chains operate today, with more ability to be flexible to what solutions are recommended. It would be great to be able to optimize a network of warehouses and delivery points but even with the optimizing computation available, the reality of what can be done might still be too much change for the supply chain as it is designed. This is because there are still fundamental physical assets (e.g. manufacturing sites, warehouses, distribution centres) required in the supply chain and for most companies, being able to change them at speed is not a realistic ask each time a model suggests a new approach. What will come from this will be the use of flexible warehousing and the role of third-party providers who own assets in key locations. They can flex their operations to support capacity and could become the key players who help dictate the future winners in supply chain.

I am most excited about the advancement in supply chain visualization. I expect, in time, to see supply chains managed through a single user interface that any user can engage with to do their role. Digital agents can support the user, do the mundane tasks and recommend courses of action. In the background, there will be systems and coding, but the user need not be skilled in how to use them because everything they do will be written back to the underlying systems. They will gain trust in what is being recommended or actioned, where explanations are readily available to support them in their role.

Final reflections

The real test is moving from the theory to practice and like all innovations, there will be test cases that work and those that don't. The US internet entrepreneur Chris Dixon has a good way of explaining it, when he says to look at what innovators are spending their weekends tinkering with, because in time, that will be the work we do in companies.[19] Think to the times of the first computer coding to now developing digital assistants and agents: we are moving towards a very different future.

What cannot be disputed is that change is integral to the supply chain going forward. Disruptions are here to stay, so how we have organized supply chains thus far will need to continue to evolve. The fashionable approach to offshoring made sense as a strategy in the late 1990s when we were globalizing[20] and the trend has stuck, but adjustments may need to focus on where we situate our people to do the roles of supply chain in the future. Many companies have embraced a 'centre of excellence' (i.e. doing specific activities in one centralized location) approach for elements of supply chain activities, such as planning, which makes sense from a cost perspective.

However, as labour costs increase globally, the discussion will be steered towards how the use of technology and data can reduce the cost. Autonomous running of key elements could result in bringing the human skill sets closer to where operations and markets are, so on-shoring or near-shoring could be more appealing again. Finally, the desire to stitch together assets to form large global multinational companies may be stalling; investors are wanting more from these companies than low-digit growth. We may enter a time when the likes of the large consumer packaged goods companies become disbanded, whereby their parts become worth more than the whole. If you look at Unilever's strategy of carving out business units such as ice cream,[21] it sees the opportunity to reinvent separate supply chains.[22]

Competition does not stand still; margin enhancement and protection need innovation and experimentation with new ways to keep costs under control. In the supply chain, we can speak confidently on technology's contribution to competitive advantage (it could be argued it has never been more necessary), but the ability to get appropriate technology investment is still a challenge. Is the technology you have running the supply chain too rigid? Is it fit for what you need today? Most importantly, what do you think you will require for the future? Then the fundamental question to ask is, 'What is holding us back?' Is it the company living quarter by quarter, with short-termism encouraged by the markets? Is it the backdrop of challenging macro conditions and political instability? Is it that our leaders see levers in other functions that will provide more benefits from technology investment than ours? Could it be that we don't have a coherent strategy that connects the supply chain to the business vision? Whatever the reason (and there are many more that could be stated), for us to get investment, we need to be able to articulate a story that links what we are doing to the impact it has with the customer and how it provides the company with a competitive advantage. The role of technology must be at the heart of this.

I see a future where a master data set exists with a layer of AI that translates this and other system data into meaningful data for analysis and decision-making. This would be available on a platform level that all users in supply chain can access. There will be tools that exist in code format in the background of a platform that orchestrates users with a unified interface that they engage with through questions (outlined in Chapter 13). I doubt this will come soon because there is so much legacy to simplify and modernize, but if we were starting from scratch, for example, with a new global company and could develop the supply chain technology differently, then this could appear more quickly.

To gain a competitive advantage, simplification and getting focus back on what the supply chain is there to do is key, serve customers at the lowest cost point with enough 'bend and bounce' to react to changing macro and micro influences.

Next generation supply chains depend on being able to bring a vision for supply chain, infused and enabled with technology to life: a generic solution off the shelf with some minor customization is not the answer. Cracking this problem by following the strategies outlined in this book, challenging the status quo and determining what technology enablement can provide you in collaboration with the other key ingredients of people, process and data will be differentiating for those who succeed and be a major contributor to providing sustainable competitive advantage in my opinion.

Endnotes

1 Starr, P (2002) The great telecom implosion, The American Prospect, Princeton University, www.princeton.edu/~starr/articles/articles02/Starr-TelecomImplosion-9-02.htm (archived at https://perma.cc/4A68-8MXE)

2 Tran, M (2002) WorldCom goes bankrupt, *The Guardian*, www.theguardian.com/world/2002/jul/22/qanda.worldcom (archived at https://perma.cc/5ZUJ-9ZH5)

3 Constable, S (2021) How the Enron scandal changed American business forever, *Time Magazine*, https://time.com/6125253/enron-scandal-changed-american-business-forever (archived at https://perma.cc/52B8-JC45)

4 Nokia (2025) History shaping society with technical innovations for 100 years, Nokia, www.nokia.com/bell-labs/about/history (archived at https://perma.cc/9BWM-5S8X)

5 Nokia (2025) Nokia Bell Labs celebrates 100 years of innovation and looks ahead to another century of discovery, Nokia, www.nokia.com/bell-labs/nokia-bell-labs-celebrates-100-years-of-innovation-and-looks-ahead-to-another-century-of-discovery (archived at https://perma.cc/9NXW-M8H7)

6 Gertner, J (2013) *The Idea Factory Bell Labs and the Great Age of American Innovation*, Penguin Books

7 Manenti, P (2025) Supply chain's future is more certain than it seems, Gartner, www.gartner.com/en/supply-chain/insights/beyond-supply-chain-blog/supply-chains-future-is-more-certain-than-it-seems (archived at https://perma.cc/5N22-L9S2)

8 Turner, C (2024) Supply chain – what is it? Understanding the backbone of modern business, Thornton & Lowe, https://thorntonandlowe.com/supply-chain-what-is-it-understanding-the-backbone-of-modern-business/#:~:text=A%20supply%20chain%20is%20a,service%20reaches%20the%20end%20consumer (archived at https://perma.cc/U8QW-DNCP)

9 Wilson, J (2024) The impact of the global supply chain crisis, Imperial College London, www.imperial.ac.uk/stories/global-supply-chain-crisis (archived at https://perma.cc/L8MP-JG6M)

10 Saili (2024) How e-commerce is driving new trends in global trade and logistics, Global Trade, www.globaltrademag.com/how-e-commerce-is-driving-new-trends-in-global-trade-and-logistics (archived at https://perma.cc/KEU2-AN58)

11 Coriddi, J (2008) An inconvenient truth, *Policy & Practice: A Development Education Review*, www.developmenteducationreview.com/issue/issue-6/inconvenient-truth (archived at https://perma.cc/B9XG-WZK8)

12 Jones, A (2025) Sustainability without the SDGs: US policy shifts and corporate ESG, Harvard Law School Forum on Corporate Governance, https://corpgov.law.harvard.edu/2025/04/03/sustainability-without-the-sdgs-us-policy-shifts-and-corporate-esg/#:~:text=In%20March%202025%2C%20the%20US,support%20for%20multilateral%20sustainability%20frameworks (archived at https://perma.cc/RS3H-N9R9)

13 Lee, Y H, Golinska-Dawson, P and Wu, J-Z (2016) Mathematical models for supply chain management, Mathematical Models for Supply Chain Management, Wiley Online Library, https://onlinelibrary.wiley.com/doi/10.1155/2016/6167290 (archived at https://perma.cc/3943-A4RL)

14 Singla, A, Sukharevsky, A, Gaffey, B, Kaka, N, Dahlström, P, Travasoni, A, Atluri, V, Lajous, T, Vieria, B and Garcia de la Torre, V (2024) Beyond the hype: Capturing the potential of AI and gen AI in tech, media, and telecom, McKinsey & Company, www.mckinsey.com/~/media/mckinsey/industries/technology%20media%20and%20telecommunications/high%20tech/our%20insights/beyond%20the%20hype%20capturing%20the%20potential%20of%20ai%20and%20gen%20ai%20in%20tmt/beyond-the-hype-capturing-the-potential-of-ai-and-gen-ai-in-tmt.pdf (archived at https://perma.cc/H4BY-E4FJ)

15 Shelley, S (2024) Navigating the digital age: The role of technology in supply chain management, University of the Cumberlands, www.ucumberlands.edu/blog/the-role-of-technology-in-supply-chain-management#:~:text=AI%20and%20machine%20learning%20are,companies%20manage%20disruptions%20more%20effectively (archived at https://perma.cc/6XWM-RKVC)

16 Jegan Joseph Jerome, J, Sonwaney, V, Bryde, D and Graham, G (2023) Achieving competitive advantage through technology-driven proactive supply chain risk management: an empirical study, Liverpool John Moores University, https://researchonline.ljmu.ac.uk/id/eprint/21563/4/Achieving%20competitive%20advantage%20through%20technology-driven%20proactive%20supply%20chain%20risk%20management%20-%20an%20empirical%20study.pdf (archived at https://perma.cc/N72J-XCED)

17 Chui, M, Collins, M and Patel, M (2021) The Internet of Things: Catching up to an accelerated opportunity, McKinsey and Company, www.mckinsey.com/~/media/mckinsey/business%20functions/mckinsey%20digital/our%20insights/iot%20valu e%20set%20to%20accelerate%20through%202030%20where%20and%20how%20to%20capture%20it/the-internet-of-things-catching-up-to-an-accelerating-opportunity-final.pdf (archived at https://perma.cc/T6JG-2EE6)

18 Wingate, D, Burns, B L and Barney, J B (2025) Why AI will not provide sustainable competitive advantage, *MIT Sloan Management Review*, https://sloanreview.mit.edu/article/why-ai-will-not-provide-sustainable-competitive-advantage/?social_token=8473eefd6fa146422d7bedef48094cd0&utm_source=linkedin&utm_medium=social&utm_campaign=sm-direct (archived at https://perma.cc/8W6Z-TMJ3)

19 Dixon, C (2024) *Read Write Own: Building the Next Era of the Internet*, Cornerstone Press

20 Crafts, N F R (2004) The world economy in the 1990s: A long run perspective, Department of Economics History, London School of Economics, https://eprints.lse.ac.uk/22334/1/WP87.pdf (archived at https://perma.cc/YHS5-7XMZ)

21 Naidu, R and Mathews, E (2024) Unilever to spin off ice cream business, cut 7,500 jobs for cost savings, Reuters, www.reuters.com/markets/deals/unilever-spin-off-ben-jerrys-unit-launches-cost-savings-plan-2024-03-19 (archived at https://perma.cc/ZBN7-NQLE)

22 Unilever PLC (2024) Growth Action Plan 2030, Unilever PLC, www.unilever.com/investors/results-presentations-webcasts/unilever-investor-event-2024/?ref=strategypunk.com (archived at https://perma.cc/PY2Q-YU8S)

ACKNOWLEDGEMENTS

When approached to write this book, it seemed a daunting task but Nick Hoar from Kogan Page helped me through the initial phases of an idea, to a synopsis and then signing off the plan for the book. Then the baton was handed to Ellen Capon who did an amazing job of being both taskmaster and cheerleader throughout the writing as my editor. Arthur Thompson guided me through the production process and Sarah Grant diligently copyedited the book. To the entire team at Kogan Page I am grateful for your patience, kindness and support through this journey.

The following people were very generous to share time and their experiences with me. You will find nuggets of their insights sprinkled through the book and it was lovely to reconnect with many of them since we rarely get the chance due to our busy schedules: Akhil Sivanandan, Anne Rees, Bart Derde, Ben Kelly, Chris Silvester, James Cordall, Johnathon Marshall, Kerryn Haynes, Lisa Callinan, Mark Hale, Matthias Gräfe, Nigel Issa and Ross Eggleton.

I am extremely grateful to Andras Csiky, Brian White, Keith Wishart and Mark Harrison for helping me with chapter reviews, notes and encouragement as well as having to turn them around quickly for me. To Jonathan Bartley for helping me early on with a challenge about structure and sending countless research items for me to consider.

I also want to thank my legal counsel Matt Dowd for his sage advice. A phone conversation in the pouring rain as I watched my eldest daughter play football was a highlight when we were discussing film rights and whether to dispute the inclusion of them in the book contract… could you imagine!

From my career to date, there are far too many people I need to thank, and by listing a few here, I know I will have forgotten or mildly offended others, so I won't – they should all know who they are.

To my 'wee' family Sarah, Saoirse, Eilidh and Morven I am grateful to you every day for making me realise there are other things to life than supply chains. To my wider family and friends, thank you for keeping me moving forward and always being available for a call at any hour.

Finally, to you the reader, thank you for taking the time to read this book.

INDEX

Note: Page numbers in *italics* refer to figures or tables.